INSTALLATION, TESTING AND MAINTENANCE

(For B.Tech./B.E./Diploma in Engineering Students)

Mohammad Yunus Khan

INSTALLATION, TESTING
AND
MAINTENANCE

Mohammad Yunus Khan
Assistant Professor,
Mechanical Engineering Section,
University Polytechnic,
Aligarh Muslim University,
Aligarh

S.K. KATARIA & SONS®
Publisher of Engineering Books
4885/109, Prakash Mahal, Dr. Subhash Bhargav Lane,
Opposite Delhi Medical Association, Daryaganj, New Delhi-110002
Phones: +91-11-43551243, Mobile: +91-9871775858, 7678425209
e-mail: katariabook@gmail.com, katariabooks@yahoo.com
Website: www.skkatariaandsons.com
(e-books, subscriptions and bulk buying also available on our website)

Published By :

S.K. KATARIA & SONS®

4885/109, Prakash Mahal, Dr. Subhash Bhargav Lane,
Opposite Delhi Medical Association, Daryaganj, New Delhi-110002
Phone: +91-11-43551243, Mobile: +91-9871775858, 7678425209
e-mail: katariabook@gmail.com, katariabooks@yahoo.com

First Edition: July, 2012, Reprint: 2026

Price: ₹ 295/-

Laser Typesetting at: Goswami Associates, Delhi.

Printed at: Printed at : N. K. Printers

Preface

This book is based mainly on the lecture notes prepared by me for classes. I want to pay my gratefulness and express my indebtness to the esteemed authors and publishers of the books used in preparing the notes. Since the matter of subject is vast, I have to concern number of books. The list of books referred, is given at the end of this book.

A diploma holder in mechanical engineering or production engineering is supposed to work as maintenance supervisor. In the capacity of a supervisor, he has to supervise installation, testing and maintenance of machines. He should be aware of safety practices used in industries. He is responsible to plan maintenance schedule and keep machines in operating condition. He has to take quick decision about replacement of machine parts. He should have good knowledge about different materials used in engineering practices.

The subject "Installation, Testing and Maintenance" has been designed to develop sufficient knowledge, which will prove advantageous in developing skills and attitude in students so that when they will go in field they may be able to discharge their duties with full confidence. It is hoped that the book will prove to be invaluable text for the readers. This book is written precisely to cater to the needs of final year diploma engineering students. Although the subject matter is very vast, I have written this book keeping in mind the syllabus of. Aligarh Muslim University (Aligarh) and Department of Technical Education (Haryana). It is hoped that this book will be helpful to students of other polytechnics and engineering colleges and also to all persons associated with maintenance work in industries.

An effort of this scope is bound to have its flaws. Constructive criticism and suggestions for the improvement of this book shall be gratefully received and appreciated.

Mohammad Yunus Khan

Acknowledgements

Al-hamd ulillah—all praises to Almighty Allah, the most beneficent and the most merciful, who endowed me with the power of mind and pen.

I feel immense pleasure in expressing my profound thanks to the persons who helped me during preparation of this book. I avail myself of this opportunity to tender my thanks to all my teachers specially Prof. Mohd. Muzammil. He spared his precious time to write foreword for this book. I would like to acknowledge my thanks to my brother, Mr. M. Yusuf Khan for his constant encouragement. It would have been difficult to write this book without the enthusiastic support and inspiration given to me by my friends Mr. Faiz H. Khan and Mr. Syed Husain Abbas Rizvi. I wish to thank Mr. S.K. Awasthi and Ms. Pradeepika Tiwari for their valuable suggestions. I would also like to thank my students. They have taught me many things related to learning and because of them I have been able to better write and revise the study materials on which this book is based.

I am extremely thankful to Mr. Sanjeev Kataria, Managing Director and all members of editorial, printing, binding and marketing division of S.K. Kataria and Sons (Publishers) for bringing out this book with high quality. Special thanks to Mr. Sharda Singh, who looked after publication work and Mr. Goswami, who helped untiringly in producing drawings.

Finally, yet importantly, I would like to express my heartfelt thanks to my parents. Words fail when I think of all, that my parents have done for me. I would like to thank them for their blessings and moral support, without which, the present work would not have seen the light of the dawn.

Mohammad Yunus Khan

Foreword

It is a matter of great pleasure and privilege for me to write this foreword for the book "Installation, Testing and Maintenance" by Mr. Mohammad Yunus Khan. Not many books on this subject are available today that comprehensively address the subject and in this context, the present book is perhaps the first of its kind and the author deserves congratulations.

The book "Installation, Testing and Maintenance" is a scholarly treatise on the subject combining simplicity with empirical know-how and visionary depth with comprehensiveness. It has been designed to develop sufficient knowledge, which will prove advantageous in developing skills among students so that when they will go in field they may be able to discharge their duties efficiently.

A mechanical engineer may work as maintenance supervisor in any industry. In that capacity, he/she faces problems of machine installation, machine testing, its failures, repairs and upkeep, day in and day out. This book, I am sure will serve as a guide to him/her. The author has tried to explore as to how maintenance engineering is basically a field of managerial and technical importance. A lot of money is spent every year in maintaining a plant. Maintenance is a key element in increasing the competitiveness of industry. The author asserts, after meticulous analysis, that if maintenance is not given its due attention, it would be disastrous. It is also essential that the maintenance supervisor should be well versed in various safety practices used in industries. The moot point is that industrial safety is basic to both machines as well as personnel. A good maintenance supervisor has to ensure this.

The book is a lucidly written text on Installation, Testing and Maintenance. The author had examined the fascinating intricacies of the subject. It is divided into twelve chapters which cover various aspects of the subject. It is designed in such a way that it will not only be useful for teachers and students but also for people working in the industry. This work is a commendable beginning in this area. The structure and content of the book will surely enhance the capabilities of the readers.

I am really impressed by the painstaking extra-ordinary efforts put-in by the author in conceptualizing and formulating this book and compliment him for this enormous work. I wish him success. I also whole-heartedly thank author who gave me the privilege of being the first reader of book and for asking me to write the foreword for the same.

I am confident that the readers will find this book as interesting and useful.

Prof. M. Muzammil
Department of Mechanical Engineering
Faculty of Engineering and Technology,
A.M.U., Aligarh

Dedicated to

My Parents

Contents

Safety Practices

CHAPTER 1

1.1 CONCEPT OF INDUSTRIAL SAFETY

Technological development is taking place at a very rapid pace in all the fields of engineering. For mobility and moving around automobiles, trains and aeroplanes are available. We have got bridges, various machineries, explosives, chemical industries etc. With this advancement of technology, chances of accidents are also increasing. Thus, the danger of human life is also more. The significance of industrial safety was realized because every year millions of industrial accidents take place, which result in either death or in temporary/permanent injury to the employees. It also involves loss of property, wastage of man, power and machine hour. Nowadays more attention is being paid in this regard.

Hazards refer to a potential condition, which might get converted into an accident. Accident may be defined as unforeseen, uncontrolled, undesirable and sudden mishap which may result in injuries or death of person, loss of property, interruptions in activities (or function) in industry. Industrial safety is mainly concerned with minimizing, controlling and eliminating hazards. Safety includes all operations and activities which reduce accidents.

Industrial safety is of great importance because if safety measures are not taken chances of accidents are definitely going to increase. Every industry appoint specially trained person as safety engineers. The government has also framed rules and regulations for industrial safety.

Some of the causes of accidents in industries are obsolete or poorly maintained equipment, unguarded machines, untested machines, violation of norms, loose clothes, not using goggles/ gloves/ masks, wearing high heel shoes, smoking in premises, human causes, environmental factors, long duration of work or duty shift, lack of knowledge etc.

1.2 SAFETY CONSCIOUSNESS

Safety consciousness is defined as the awareness and carefulness of the workers to prevent the accidents. While at work, the workers must always be fully aware of accidents that may take place and must take all necessary precautions for their safety. Safety consciousness can be increased by adopting the given methods:

(*i*) Forming safety committees.

(*ii*) Organizing safety awareness programs for the workers.

(*iii*) Sending related information and literature to the worker.
(*iv*) Displaying safety posters and boards to remind the workers about safety.
(*v*) Training workers to use safety devices.
(*vi*) Identifying all accident prone areas.
(*vii*) Welcoming all important suggestions related to safety.

1.3 SAFETY IN HANDLING OF HOISTING EQUIPMENTS AND CONVEYORS

Hoists are used for either lifting or lowering the loads. Conveyors are used for moving the material either vertically or horizontally between fixed stations. A high degree of safety is required while operating these equipments. Few important precautions related to handling of hoisting equipments are as follows:

(*i*) Hoisting equipments should be of good material, proper mechanical construction, high strength and free from defects.
(*ii*) It should be operated by fully trained worker. Worker should be mentally, physically and emotionally fit.
(*iii*) The control cord of hoist should carry distinct marking 'hoist' or 'lower' to avoid unnecessary confusion and should be of non-conducting material.
(*iv*) The control panel should be provided with thick rubber or plastic matting (insulation) to prevent chances of electric shock to the operator.
(*v*) Regular inspection should be carried out and regular maintenance schedule should be followed.
(*vi*) Repair and maintenance work should be carried out by trained person. Such work should be carried out only when equipment is switched off.
(*vii*) Standard rules for operation should be strictly followed.
(*viii*) Worker engaged in arranging loads or otherwise working near the hoisting equipment should never be allowed to come under the loads.
(*ix*) Protection against fire or explosion are needed when diesel operated equipments are being used.
(*x*) Mobile cranes should not be used for suspended load except for low loads and that to for short distances.

Few important precautions related to handling of conveyors are as follows:

(*i*) All rotating parts (like sprockets, gears etc.) should be properly covered.
(*ii*) Effective measures should be taken so that the safe working speed is not exceeded.
(*iii*) An effective lubrication schedule should be worked out and implemented.
(*iv*) Regular inspection should be carried out and worn out parts, if any, should be replaced immediately.
(*v*) Repair and maintenance work should be carried out by trained person.
(*vi*) Proper load signals should be displayed on all loading stations.
(*vii*) Riding on a conveyor by worker should be strictly prohibited.
(*viii*) All the persons working on or near the conveyor should wear tight fitting clothes.
(*ix*) Samples should not be collected when conveyors are moving (especially if bucket conveyor is used).

(*x*) Sufficient lighting, ventilation, drainage and escape ways should be provided when conveyors are to be used in pits, tunnels and similar other enclosures.

1.4 PROTECTION IN STORAGE

1.4.1 Fire Protection

(*i*) Employees should be sensitized and made aware of hazard of accidents and carelessness on their parts.

(*ii*) "No Smoking" signs should be prominently displayed and strictly observed especially where highly inflammable substances are stored.

(*iii*) Adequate fire protecting equipments such as high-pressure water lines, water sprinkler system, chemical extinguishers, fire pads, sand should be provided.

(*iv*) Goods should be stored in fire proof container and building. Petrol and diesel should be stored in underground tanks. Explosives should be stored at a distance from all other goods.

1.4.2 Protection from Pollutants

(*i*) Goods should be in boxes or closed cabinets.

(*ii*) Air-conditioned storage may be provided for materials like paper, textile and electronic equipments.

(*iii*) Exhaust fans should be used to draw out contaminants, if any.

1.4.3 Protection against Extreme Weather Conditions

(*i*) There are some materials whose nature requires that they should be stored outdoors. They should be protected by sheds, duck bag or plastic covering.

(*ii*) There are certain goods which should be stored in moisture free space especially electronic items.

1.4.4 Protection against Theft

(*i*) Protection against theft can be ensured by properly locking the store room.

(*ii*) All persons who are not related with the issues of stores should be excluded.

(*iii*) Goods should be issued only on authority of properly signed requisition.

1.4.5 Protection against Deterioration

(*i*) Some material deteriorates with passage of time. They should not be overstocked and 'first-in-first-out' storage system should be adopted for such material.

(*ii*) Another form of deterioration is the damage which may result from improper material handling. Proper material handling system should be adopted.

1.5 PROTECTION IN MANUAL HANDLING OF MATERIAL

Material handling is very common functions in any industry. Every worker is involved in material handling at one stage or another. Some are exclusively meant for this purpose while others are required to perform these functions as a part of their manufacturing activities. These functions may be performed either manually or by using mechanical devices. If they are performed mechanically, the chance of injuries/damages is comparatively less.

Some of the important precautions for manual handling of materials are:

(*i*) Workers should be fully trained in the proper ways of lifting and setting down the objects.

(*ii*) The wet and dirty objects, with slippery surfaces (greasy or oily surface) should be properly cleaned before handling. The hands of the worker should also be free of oil or grease.

(*iii*) To prevent hand injuries, the worker should wear protective clothing like leather hand gloves and sleeves.

(*iv*) In order to prevent foot injuries, the worker should wear foot protection garments (shoes). Many a time, the type of object to be handled may require the workers to wear protective shield for protecting their eyes and heads.

(*v*) Drums and barrels should be handled using some kind of mechanical lifting and carrying device.

(*vi*) While carrying object to some distance, it should be seen that the path is free and not slippery.

(*vii*) The workers should be properly trained for keeping correct back and knee positions, hold the object close to the body while lifting and carrying, correct and tight grip should be applied, position of chin and application of body weight in lifting and settling down by hand should be correct. This will prevent muscle strains and back injuries to the workers.

(*viii*) When a group of workers carry a heavy load, their lifting, walking and settling down actions should synchronize.

1.6 SAFETY IN HANDLING OF CHEMICALS

Few important precautions related to handling of chemicals are as follows:

(*i*) The chemical laboratory should be kept clean and in proper order.

(*ii*) Volatile chemicals should not be stored near source of heat or directly under sunlight.

(*iii*) Bottles of chemicals should be properly labeled.

(*iv*) Hands should be protected with rubber gloves while handling dangerous chemicals.

(*v*) Waste chemicals and broken glass should be disposed off at correct place.

(*vi*) Bottles containing acids should not be carried in the hand. For this purpose, bottle carrier should be used.

(*vii*) Laboratory should have proper ventilation arrangement.

(*viii*) Fire extinguishers should be available in the laboratory.

1.7 SAFETY PRACTICES IN WORKSHOP

The statement "accidents do not take place, they are caused" should be kept in mind and safety practices should be strictly followed in workshop. The most important tool in any shop is human brain and it should be used. Safety practices include all those operations and activities which reduce accidents. There are number of precautions to be followed for protection against various hazards.

Following are some general precautions that should be kept in mind while working in workshop:

(*i*) Always wear apron and shoes.

(*ii*) Wear safety goggles when performing welding operation and in area where chips of wood and metal are flying.

(*iii*) Avoid wearing loose clothing that might get caught in rotating parts. Also, avoid wearing neck tie, jewellery, ring, watch and bracelet while working in workshop.

(*iv*) Keep the working area clean. Oil, gas and other leakages should be checked. All spills, sawdust, wood and metal chips should be cleaned immediately.

(*v*) All the machine parts should be properly guarded.

(*vi*) Machine should be operated by trained person only.

(*vii*) Use correct tools and use tools correctly.

(*viii*) Tools should be stored properly and at specific place.

(*ix*) Electrical connections should be proper and proper earthing should be provided.

(*x*) Never distract or interface with any worker, operating machine.

(*xi*) Maintain proper housekeeping.

(*xii*) Any defective equipment/tool or improperly guarded machine parts should be reported to safety supervisor.

In the following headings, safety practices to be followed in different shops are discussed.

1.8 SAFETY IN CARPENTRY SHOP

Carpentry includes making joints of wood for different applications like door, window, furniture and packing etc. While working in carpentry shop, following precautions should be followed:

(*i*) Always wear apron, shoes and safety goggles. Avoid wearing loose clothing.

(*ii*) Keep the working area clean. Saw dust, nails should be immediately cleaned.

(*iii*) Sharp tools should be carefully used.

(*iv*) If saw is not cutting properly then don't force it. Check what is wrong.

(*v*) Files and chisels used should be fitted proper handle.

(*vi*) Force should be applied only in the forward cutting stroke relieve the force in the backward stroke.

(*vii*) Job/workpiece should be tightly secured properly in vice.

(*viii*) Tools should be stored properly and at specific place.

(*ix*) Never place unguarded cutting tools at improper place like drawer. It can cause severe hand injuries.

(*x*) There are varieties of power saws used in carpentry shop like band saws, circular saws, jig saw etc. Following safety guidelines should be followed while working on band saw:

(*a*) Blade of band saw should be set with the proper amount of tension.

(*b*) Worker should keep his hands on either side of the cutting line.

(*c*) Radius of cutting area should not be too small for the saw blade.

Following safety guidelines should be followed while working on circular saw:

(*a*) Never try to clean scraps with fingers.

(*b*) Thin tubular material should not be cut with a circular saw.

(*c*) Special care should be taken while cutting wood with twisted lumber.

1.9 SAFETY IN FITTING SHOP

Following are some of the precautions that should be followed while working in fitting shop:

(*i*) Always wear apron and shoes. Avoid wearing loose clothing.

(*ii*) Avoid wearing neck tie, jewellery, ring, watch and bracelet while carrying fitting work.

(*iii*) Job/workpiece should be tightly secured properly in vice.

(*iv*) Sharp tools should be carefully used. Playing with tools should be strictly avoided.

(*v*) Files and chisels used should be fitted proper handle.

(*vi*) Force should be applied only in the forward cutting stroke relieve the force in the backward stroke.

(*vii*) Never use a chisel or knife as screwdriver.

(*viii*) Never carry sharp tools in pocket.

(*ix*) Blade of hacksaw must be checked for proper fitting in pins.

(*x*) When working with chisel, material must be chipped away.

1.10 SAFETY PRACTICES IN WELDING

The welding is a process of joining two similar or dissimilar metals by fusion, with or without the application of pressure and with or without the use of filler material. The fusion of metal takes place by means of heat, which may be obtained from the blacksmith's fire, electric arc, electric resistance or by chemical reaction. Some general precautions that should be followed while carrying out welding operation are as given below:

(*i*) Welding operation should be carried out by well-trained welders.

(*ii*) Welders must wear welding hoods and shields, goggles, gloves and apron. Otherwise injuries like burns, damage to eye, electric shock, cuts etc can take place. It only takes few second for radiation to cause damage to the cornea as well as burns to the eye and skin of the welder.

(*iii*) Welders should wear appropriate dress. The clothes should be free from oil, grease or dampness. Wearing synthetic fibre during welding can be dangerous.

(*iv*) Place of welding should be safe and free from combustible substances. It should also be free from chemicals and grease.

(*v*) Place of welding should be equipped with fire extinguishers.

(*vi*) Welding operation on different jobs should be carried out on suitably designed tables.

(*vii*) Welding operation should not be carried out on any vessel which is filled with inflammable or explosive substance. If welding operation is to be carried out on such vessel, then guards must be used to confine the heat, sparks and slag to protect fire hazards.

(*viii*) One of the most common injuries in welding results from general burns by touching hot surfaces. Thus, touching of hot surfaces must be avoided.

1.10.1 Safety Practices in Electric Welding

Some of the safety practices in electric welding are as discussed below:

(*i*) All cables and connections should be checked before starting welding operation.

(*ii*) Unidentified electrode or electrode material should never be used.

(*iii*) Welding cables and connections should be in proper conditions. Cable should be continuous and properly insulated.

(*iv*) Earthing arrangement should be proper.

(*v*) Outer surfaces of electrode holder should be properly insulated.

(*vi*) Electrode holder when not in use should be placed in a well-insulated hook.

(vii) Wrench or similar tools should never be used for removing electrode.

(viii) Care must be taken to keep welding cable away from power cables.

1.10.2 Safety Practices in Gas Welding

Some of the safety practices in gas welding are as discussed below:

(i) All connections of gas cylinders should be checked before starting gas welding operation. Regulators, blow pipes and torch should be checked carefully before starting welding.

(ii) Smoking by the welder should be strictly prohibited when carrying out welding operation.

(iii) Cylinder valve, regulators and torches should be kept cleaned.

(iv) Valves should be closed when work is completed or stopped for some time. Also, they should be closed before transporting cylinder to other place.

(v) Standard keys or spanners should be used for opening and closing valves of gas cylinders.

(vi) Proper technique should be employed to detect leakage of gases.

(vii) Oxygen gas cylinder valves and fittings should not be handled with oily or greasy hands.

(viii) Acetylene gas cylinder and its accessories should not be fitted with parts made up of copper alloys.

1.11 SAFETY PRACTICES IN FOUNDRY SHOP

Foundry shop is a source of many types of hazards. Some of the safety guidelines to be followed while working in foundry shop are discussed below:

(i) Always wear apron, gloves, leather shoes, face shield and hat. Avoid wearing loose clothing. Safety glass is also very essential. Any negligence can cause blindness.

(ii) Water should not be intruded in the furnace or crucible. Even a trace of water can cause a large explosion.

(iii) Molten metal should not be poured on wet ground. It can cause explosion.

(iv) Water should not be used on metal fire. Dry sand should be used.

(v) Clean metal should be used as for casting purposes. Using painted metal or metal covered with lubricants can be very toxic.

(vi) Foundry shop should be kept clean and free from combustible substances.

(vii) Damaged crucible should never be used.

(viii) Foundry equipment should be inspected regularly for cracks and wear.

(ix) Foundry shop should be equipment with proper fire extinguishers.

(x) Foundry should be well-ventilated. Fumes and gases generated during combustion and other chemicals used can be very fatal. Molding sand contains silica. Silica dust exposure can cause various lung diseases.

1.12 SAFETY PRACTICES IN SMITHY SHOP

The following safety guidelines must be followed while working in smithy shop:

(i) Hammers used in forging shop should have proper handle.

(ii) Blunt chisels should not be used.

(iii) Heavy blows should not be given on anvil tail.

(*iv*) Proper length tong should be used.

(*v*) The tong should be kept cool while working with it.

(*vi*) Fullers and swages should not be used for cold metal.

(*vii*) Hearth should be ignited properly. Fire should remain free from clinkers.

(*viii*) While working on furnace, fire should be kept small but deep.

(*ix*) Hammering work should not be carried out in lazy mood.

(*x*) Fresh coke should not be added on the top of fire. Instead, it should be added on side and utilized as per requirements.

1.13 SAFETY PRACTICES IN MACHINE SHOP

There can be number of ways in which injuries to the worker can take place in machine shop. The point of danger in various machines can be movement of rotating members, reciprocating arms, moving belts, meshing gears, cutting teeth, punching, boring, shaping etc. The points of danger between belt and pulley, chain and sprocket, and rack and pinion are shown in Figure 1.1. The dangerous points in meshing gears and rolling mills are as shown in Figure 1.2. Point of hazards in screw conveyors, flywheel and the periphery of grinding wheel are shown in Figure 1.3. The cutting hazards in circular saws, drilling machines, lathe machine and milling machines are shown in Figure 1.4.

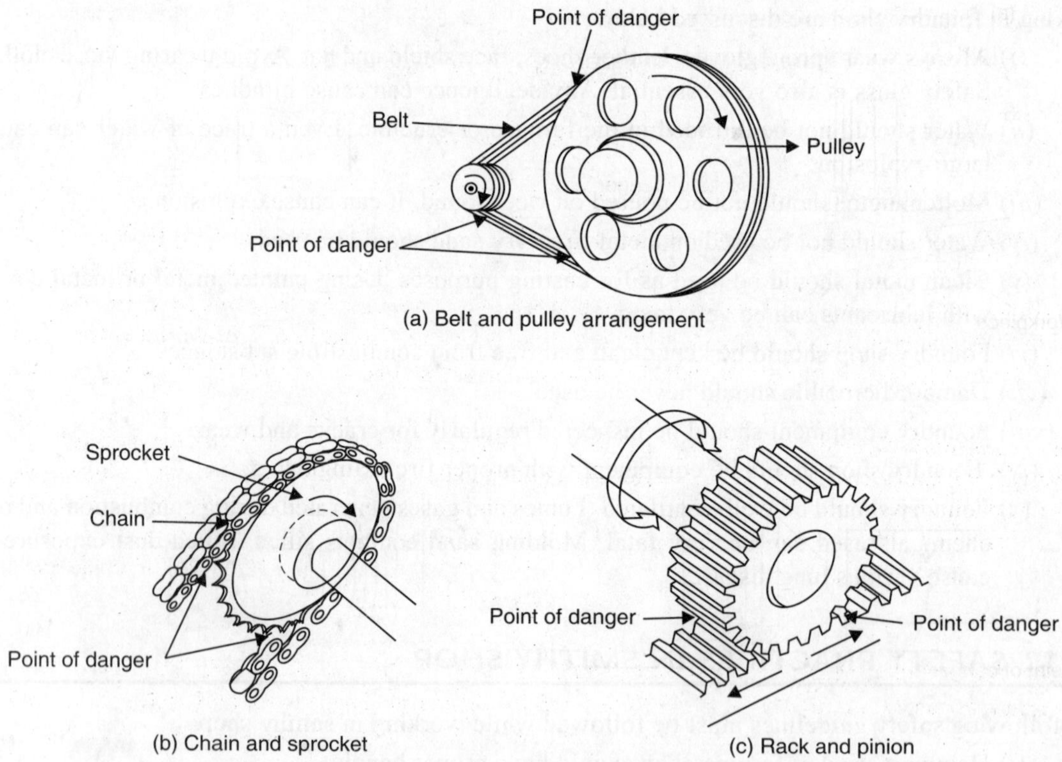

Figure 1.1 Point of danger between rotating elements

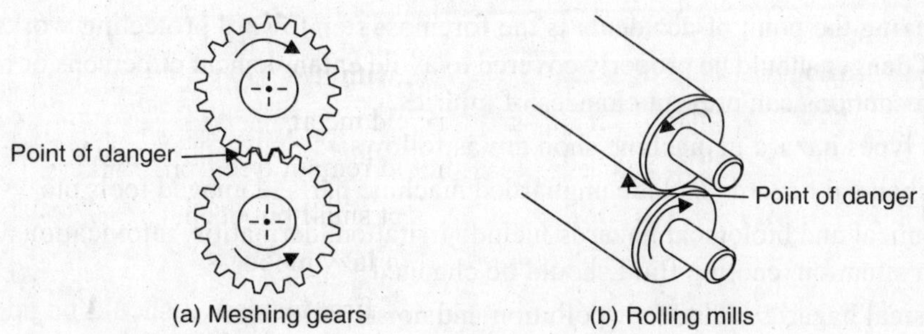

Figure 1.2 Dangerous points between some other rotating elements

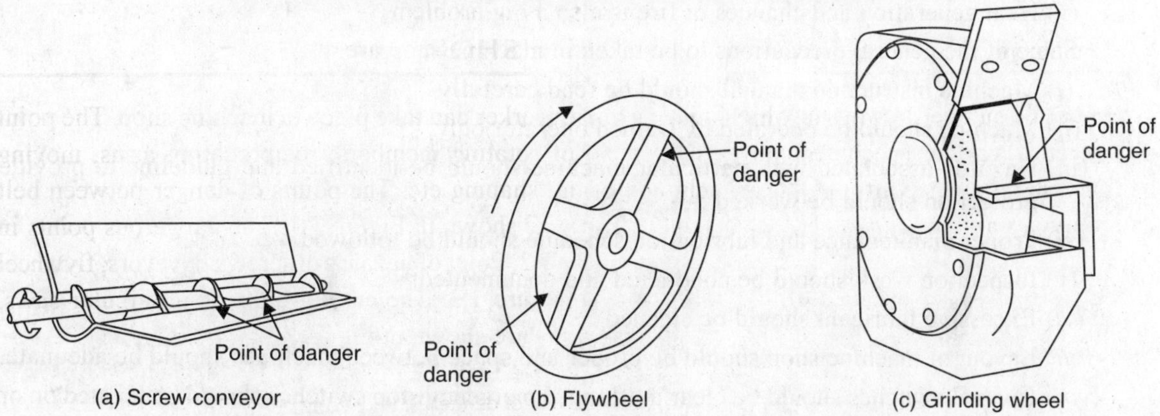

Figure 1.3 Point of hazards between rotating machine parts

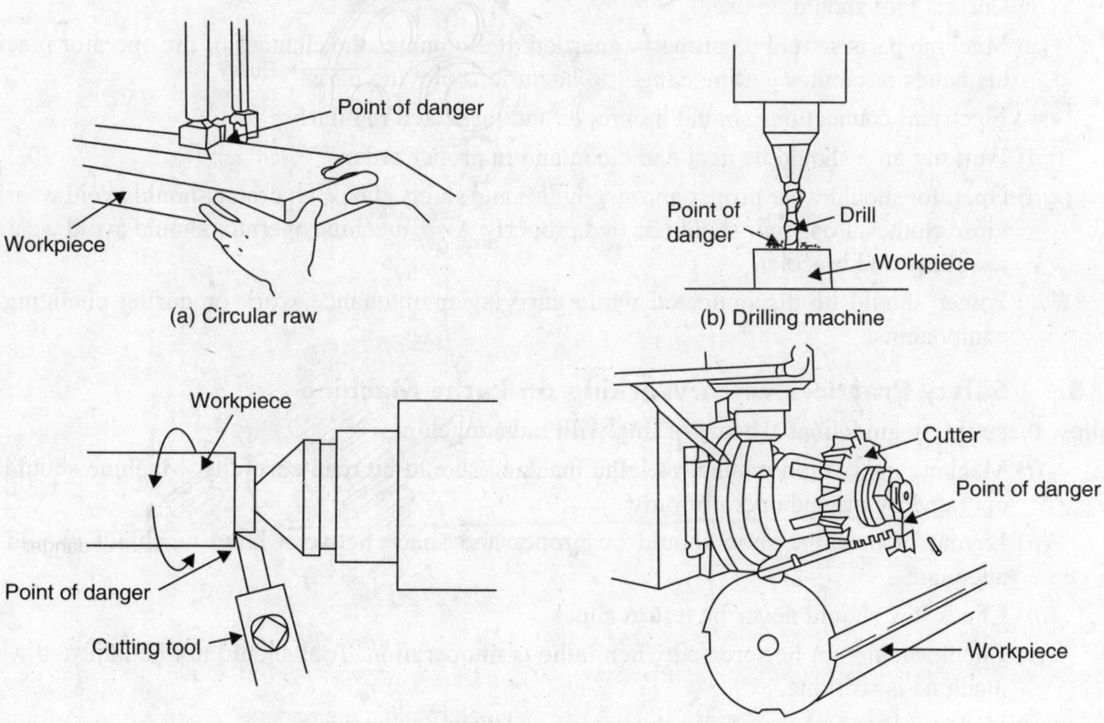

Figure 1.4 Mechanisms involving cutting hazards

Recognizing the point of accidents is the foremost step toward protecting workers from injury. These points of danger should be properly covered to avoid entanglement of persons or material. Proper training and precautions can prevent chances of injuries.

Various types hazard in machine shop are as follows:

(*i*) Mechanical hazards include unguarded machine parts, damaged tools etc.

(*ii*) Chemical and biological hazards include irritation, dermatitis, intoxication and infection. In such situation, cooling fluid should be changed.

(*iii*) Physical hazards include air pollution and noise pollution.

(*iv*) Electrical hazards include faulty connections, improper earthing etc.

(*v*) Heat generation and chances of fire is also a big problem.

Some of the general precautions to be taken in machine shop are:

(*i*) Machine instruction manual should be read carefully.

(*ii*) Machine should be operated by trained operator only.

(*iii*) Hazards associated with particular machine should be identified and guideline to provide protection should be worked out.

(*iv*) Proper maintenance and lubrication schedule should be followed.

(*v*) Inspection work should be conducted and documented.

(*vi*) Excessive lubricant should be cleaned.

(*vii*) Layout of machine shop should be proper and space between machines should be adequate.

(*viii*) Control switches should be clear marking. Emergency stop switches should be located on or near the machine.

(*ix*) Correct tool should be used.

(*x*) Machine parts should be properly guarded. It eliminates the chances of the operator placing his hands or clothes getting caught in hazardous moving parts.

(*xi*) Electrical connections should be proper and inspected regularly.

(*xii*) Working area should be neat and clean and in proper order.

(*xiii*) Operator should wear proper apron, goggles and safety shoes. Operator should avoid wearing loose clothes. Long hair should be tied properly. Also, machine operator should avoid wearing jewellery and bracelet.

(*xiv*) Power should be disconnected while carrying maintenance work or during changing of components.

1.13.1 Safety Practices when Working on Lathe Machine

Follow these safety guidelines when working with lathe machine:

(*i*) Machine instruction manual of lathe machine should be read carefully. Machine should be operated by trained operator only.

(*ii*) Layout of machine shop should be proper and space between lathe machines should be adequate.

(*iii*) Chuck key should never be left in chuck.

(*iv*) Chuck should not be screwed when lathe is in operation. Tool should not be adjusted when machine is working.

(*v*) Hands should be kept off chuck rims when lathe is operating.

(*vi*) Machine parts should be properly guarded.

(*vii*) Metal chips should never be removed by hands. Brush must be used for this purpose.

(*viii*) Machine parts in motion should not be touched. These parts should not be tightened during operation.

(*ix*) Maintenance work should be carried out by trained person. Such work should be carried out only when machine is switched off.

(*x*) Electrical connections should be proper and inspected regularly.

(*xi*) Turning work on workpiece must be started at the lowest speed. Before advancing for higher speed, it should be ensured that job is in cylindrical shape.

(*xii*) Persons not related to operation should be strictly kept away from operating machine.

1.13.2 Safety Practices when Working on Drilling Machine

Follow these safety guidelines when working on drilling machine:

(*i*) Machine instruction manual of drilling machine should be read carefully. Machine should be operated by trained operator only.

(*ii*) Machine parts should be properly covered.

(*iii*) Workpiece should be secured properly before starting drilling.

(*iv*) Center punch should be used for marking workpiece before starting drilling.

(*v*) Parts of Drill machine in operation should not be touched by hands. Also, parts should not be tightened during operation.

(*vi*) Drilling process should be carried out at correct speed. Drill should not be forced into the job.

(*vii*) Drill should be frequently backed out of cut to clean and cool drill bit.

(*viii*) Electrical connections should be proper and inspected regularly.

(*ix*) Maintenance work should be carried out by trained person. Such work should be carried out only when machine is switched off.

(*x*) Persons not related to operation should be strictly kept away from operating machine.

1.13.3 Safety Practices when Working on Grinder

Follow these safety guidelines when working on Grinder:

(*i*) Working area should be kept neat and clean. Flammable substances should be stored at a safe distance from the place where grinding operation is done. As sparks generated during grinding operation can ignite it.

(*ii*) Operator should wear proper apron and goggles for eye protection.

(*iii*) Before starting the machine, grinding wheel should be inspected. Broken or cracked wheel should never be used.

(*iv*) Operator should protect fingers from coming in contact with grinding wheel. He should wear gloves for protecting his hands.

(*v*) Components of grinder should not be adjusted when it is in operation. All rotating parts should be properly guarded.

(*vi*) Vice-grip pliers or clamp should be used to hold workpiece.

(*vii*) Operator working on grinder should avoid wearing loose clothes. Long hair, tie and scarf should be tied properly Also, operator should avoid wearing jewellery and bracelet.

(*viii*) Grinder should be left unattended when it is in operation.

(*ix*) Electrical connections should be proper and inspected regularly.

(*x*) New wheel should be installed in accordance with manufacturer's guidelines and should be operated at recommended speed. Mounting, care and inspection of grinding wheel must be carried out by competent person.

SUMMARY

- Accident may be defined as unforeseen, uncontrolled, undesirable and sudden mishap which may result in injuries or death of person, loss of property, interruptions in activities.
- Industrial safety is mainly concerned with minimizing, controlling and eliminating hazards responsible for accidents.
- Some of the causes of accidents in industries are obsolete or poorly maintained equipment, unguarded machines, untested machines, violation of norms, loose clothes, not using goggles/ gloves/ masks, smoking in premises, human causes, environmental factors, long duration of work or duty shift, lack of knowledge etc.
- Safety consciousness is defined as the awareness and carefulness of the workers to prevent the accidents.
- Regular inspection and maintenance of hoisting equipments and conveyors should be carried out by trained person.
- Employees should be made aware of fire hazards. Adequate fire protecting equipments should be available in stores.
- Air-conditioned storage is required for materials like textile and electronic equipments.
- Materials to be stored outdoor should be protected by proper covering.
- Goods from store should be issued only on authority of properly signed requisition.
- The workers should be properly trained for keeping correct back and knee positions, hold the object close to the body while lifting and carrying, correct and tight grip should be applied, position of chin and application of body weight in lifting and settling down by hand should be correct. This will prevent muscle strains and back injuries to the workers.
- The chemical laboratory should be kept clean, in proper order and well ventilated. Fire extinguishers should be available in the laboratory.
- In workshop, operator should always wear apron and shoes. Safety goggles should be worn while performing welding operation and in area where chips metal may fly.
- Loose clothing, neck tie, jewellery and bracelet should not be worn while working in workshop.
- Working area in workshop should be neat and clean and in proper order. Workshop should be equipped with fire extinguishers.
- Correct tools should be used and tools should be used correctly.
- Machine parts should be properly guarded.
- Machine instruction manual should be read carefully. Machine should be operated by trained person only.
- Layout of machine shop should be proper and space between machines should be adequate.
- Metal chips should never be removed by hands. Brush should be used for this purpose.

- Machine parts in motion should not be touched. These parts should not be tightened during operation.
- Flammable substances should be stored at a safe distance from the place where grinding operation is done.

ASSIGNMENT

1. Students should be made aware of laws and safety practices that are followed in industries.

EXERCISES

1. What is the significance of the subject "Installation, Testing and Maintenance"?
2. What is meant by safety consciousness? How it can be increased in industries?
3. List the common safety measures required to be taken while handling the hoisting equipment.
4. What general safety precautions are necessary while handling the conveyors?
5. What common safety precautions are necessary in storing of material?
6. What common safety precautions are necessary in manual handling of material?
7. Mention important precautions related to handling of chemicals.
8. What are the general precautions to be kept in mind while working in workshop?
9. What precautions are to be followed in carpentry shop?
10. Write safety measures to be followed while working on band saw and circular saw.
11. List precautions that should be followed while working in fitting shop.
12. What general precautions should be taken while carrying out welding operation?
13. Discuss safety practices in electric welding and gas welding.
14. Write notes on safety practices in foundry shop.
15. Discuss smithy shop safety practices.
16. What are various types hazard in machine shop?
17. What are the safety precautions to be observed in machine shop?
18. What are the precautions to be taken while working with lathe machine?
19. Discuss safety guidelines to be followed when working on drilling machine.
20. What are the safety precautions to be followed when working on grinder?

❑❑❑

System of Limits and Fits

2

CHAPTER

2.1 INTERCHANGEABLE MANUFACTURE AND MASS PRODUCTION

It is not possible to achieve 100% accuracy even with advance machine tools. Reasons for inaccuracies can be temperature changes, deflections in machine tool, tool wear, vibrations in machine tool and work, and human error. If the dimensions are to be maintained upto a high standard of accuracy, lot of time is consumed causing increased cost and even then there is scope for small variations. In mass production, where the work has to be done in a set competitive time, greater variations are possible. In recognition of this situation, certain variations are allowed in the sizes of the machine elements or parts. The system of manufacture in which dimensions of a part lie within certain specified limits causes interchangeable manufacture. The term interchangeable manufacture means that the parts which go into assembly may be selected at random from large number of parts. These parts have their dimensions within certain specified limits.

The interchangeable manufacture has the following benefits:

(*i*) Instead of fitting components together, which requires adjustment of size and a high degree of skill, components can be assembled.

(*ii*) The components will assemble together at random. This results in reduction in cost of production and time to build the products.

(*iii*) An assembly can be easily serviced by replacing defective parts with another components manufactured within the same range of dimensions.

(*iv*) Components can be manufactured in large batches.

(*v*) With advance machine tools, mass production of the component is fast using cheaper labour.

2.2 SYSTEM OF LIMITS AND FITS

Figure 2.1 explains the terminologies used in defining tolerance and limit. The definition of the various terms is given below. For the convenience, shaft and hole are chosen as two mating components.

1. **Shaft:** It refers not only to the diameter of circular shaft but also to any external dimension of a component.
2. **Hole:** It refers not only to the diameter of a circular hole but also to any internal dimension of a component.

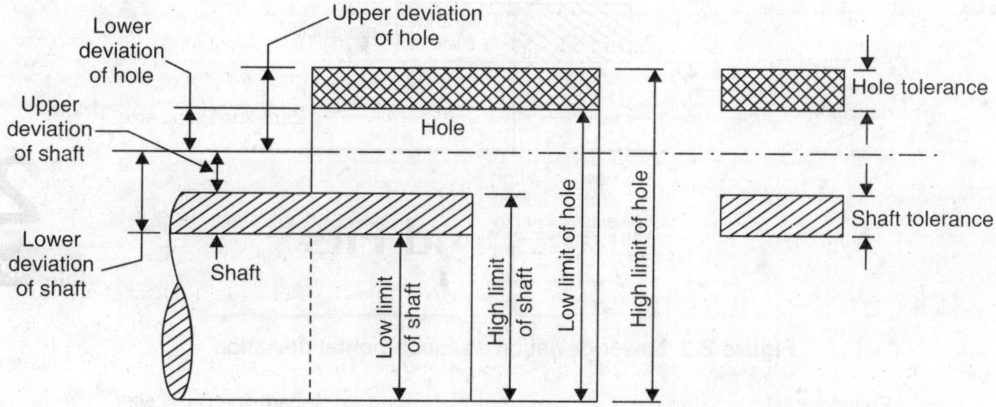

Figure 2.1 Conventional diagram of Limits

3. **Basic size:** It is the standard size for the part and is same for the hole and its shaft. This is the size which is obtained by calculation for strength (design consideration).
4. **Actual size:** It is the dimension as measured on a manufactured part (shaft or hole).
5. **Limits of size:** These are the maximum and minimum permissible sizes of the part.
   - **Maximum limit:** The maximum size permitted for the part is known as maximum limit or high limit (denoted by H.L.).
   - **Minimum limit:** The minimum size permitted for the part is known as minimum limit or low limit (denoted by L.L.).
6. **Tolerance:** It is the difference between maximum and minimum limits of the component, *i.e.*, between high limit and low limit. Depending on the application, the permissible variation of dimension is set as per available standard grades.
7. **Allowance:** It is an intentional difference between the maximum material limit of mating part. For shaft, the maximum material limit will be its high limit and for hole, it's low limit.
8. **Zero Line:** It is a line of zero deviation and represents the basic size.
9. **Deviation:** It is the algebraic difference between a size (actual size, limits of size) and the corresponding basic size.
10. **Actual Deviation:** It is the algebraic difference between an actual size and corresponding basic size.
11. **Lower Deviation:** It is the algebraic difference between the minimum limit of size and the corresponding basic size. It may be positive or negative depending upon the minimum limit of size being greater or smaller than the basic size.
12. **Upper Deviation:** It is the algebraic difference between the maximum limit of size and the corresponding basic size. It may be positive or negative depending upon the maximum limit of size being greater or smaller than the basic size.
13. **Fundamental Deviation:** It is either the upper or lower deviation which is the nearest one to the zero line for either a hole or shaft. It defines the location of the tolerance zone with respect to the zero line.

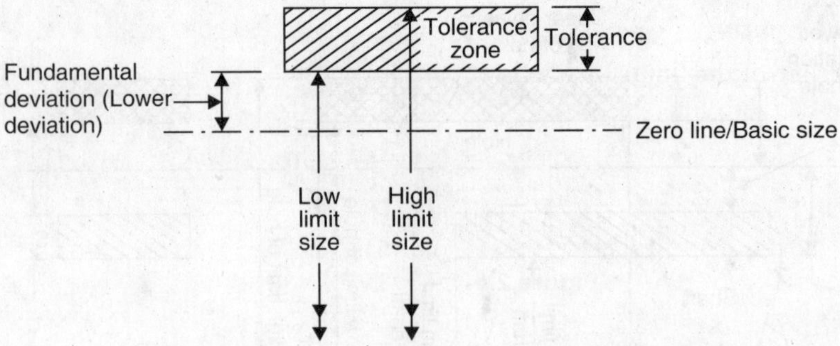

Figure 2.2 Lower deviation as fundamental deviation

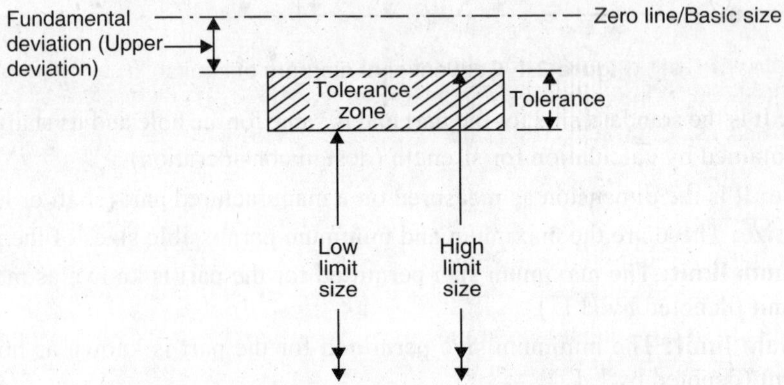

Figure 2.3 Upper deviation as fundamental deviation

14. **Tolerance Zone:** It is the zone bounded by the two limits of size of a part in the graphical presentation of tolerance. It is defined by its magnitude and by its position in relation to the zero line.
15. **Basic Shaft and Basic Hole**: The shafts and holes that have zero fundamental deviations are respectively termed as basic shaft and basic hole. The basic hole has zero lower deviation while the basic shaft has zero upper deviation.

2.3 SYSTEM OF WRITING TOLERANCE

2.3.1 Unilateral System

In unilateral system, both the limits of size are on the same side of the zero line. One of the limits of size may be the basic size.

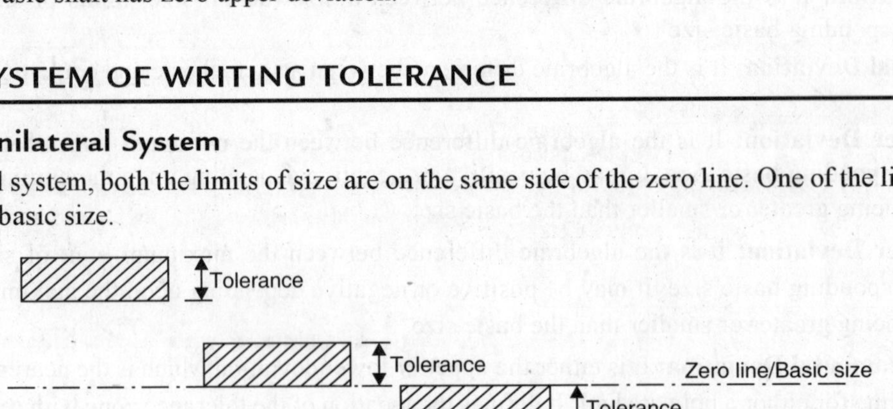

Figure 2.4 Unilateral System

2.3.2 Bilateral System

In bilateral system, one of the limits of size is on one side of the zero line and the other limit of size is on the other side of the zero line.

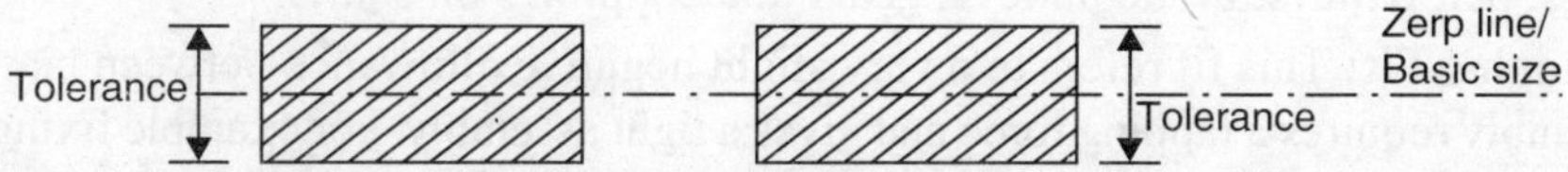

Figure 2.5 Bilateral System

2.4 FITS

Fits means a degree of tightness or looseness between two mating parts. When two mating parts fit with each other, the nature of fit is dependent on the limits of tolerances and fundamental deviations of the mating parts. On the basis of positive, zero and negative values of the clearance, three major types of fits have been defined as: Clearance fit, Transition fit and Interference fit. These have been further sub-classified as:

2.4.1 Clearance Fit

The internal member fits in an external member and always leaves clearance between the parts (*i.e.*, size of hole is always larger than the size of shaft).

(*a*) **Loose Fit:** A fit with enough clearance which allows free play of the mating part is called as loose fit. It is employed between those mating parts, where no precision is required as for example in agricultural and farm machinery, hand trolleys, plumber blocks bearing and loose pulley.

(*b*) **Free Fit:** A free fit provides less allowance and is used between mating parts, where accuracy is not essential or where large variations in temperature may occur as for example in bearing of oil engines, automobile parts, electric motors and generators etc.

(*c*) **Medium Fit:** It offers minimum allowance between two mating parts and hence, offers great precision in the movement of parts. It is used on machine tools slides, jigs, bushes, feed movement of the spindle quill in a drilling machine, piston and slide valves etc.

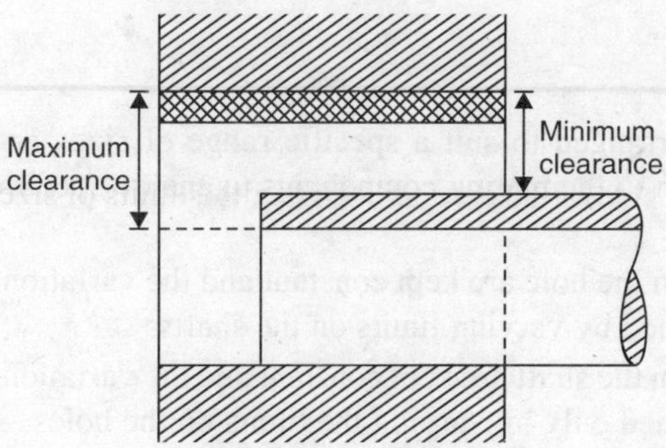

Figure 2.6(*a*) Clearance Fit

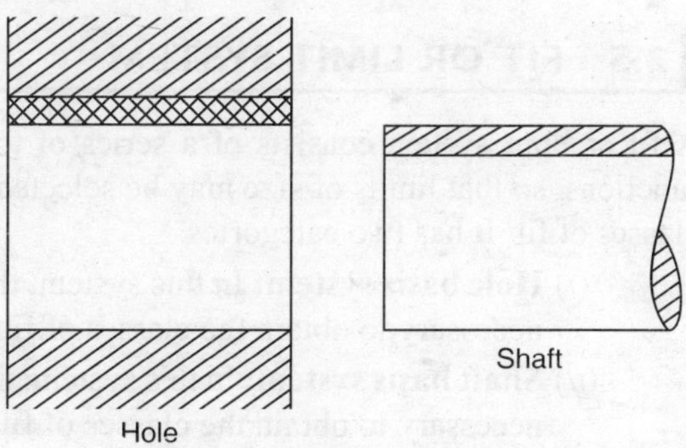

Figure 2.6(*b*) Transition Fit

2.4.2 Transition Fit

It may result in either a clearance or interference condition (*i.e.*, it lies between clearance and interference fit). In other words, we can say that tolerance zone of shaft and hole intersect.

(*a*) **Snug Fit:** This fit refers to zero allowance between mating parts and requires a very careful hand assembly. Thus, a hand-tight fit between mating parts is obtained. This type of fit is also known as push fit. The mating parts show least vibrations. Such a fit is used for locating plugs, belt pulleys, brake pulleys, gears and couplings on shafts.

(*b*) **Wringing Fit:** This fit refers to a very slight negative allowance between mating parts. This assembly requires a tapping force and gives a tight assembly. For example fixing keys, dowels, pins, gears and bearing bushes etc.

2.4.3 Interference Fit

The internal member is larger than the external member such that there is always an actual interference of material (*i.e.*, size of hole is always smaller than the size of shaft).

(*a*) **Tight Fit:** A fit between mating parts with slight negative allowance, requiring light to moderate force to assemble. The negative allowance is larger than the wringing fit. It is used such cases where parts are to be maintained in position during running conditions such as pulley keyed to the shaft, rocker arm etc.

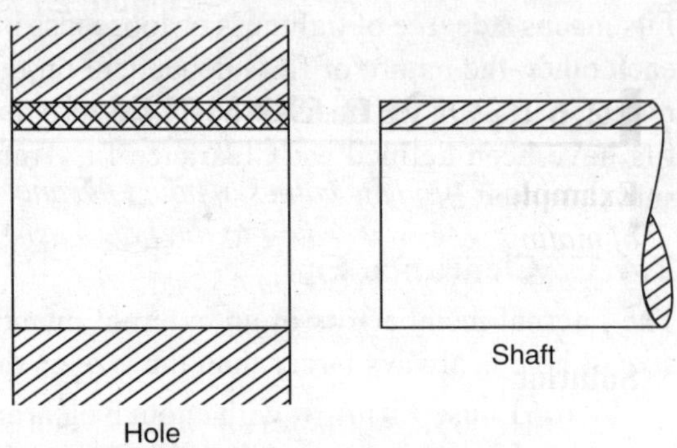

Figure 2.6(*c*) Interference Fit

(*b*) **Medium Force Fit:** The negative allowance is fit appreciable and therefore, greater pressure is required for assembly. Such fit is assembled through the use of a press or hand hammering. The fit offers a permanent type of assembly. For example, stepped pulley on the driver shaft.

(*c*) **Heavy Force Fit:** It refers to maximum negative allowance. Heavy force is need for the assembly. The fitting of frame or rim can also be obtained by heating the rim and then rapidly cooling it, in its position. Such fits are not to be disassembled during their entire service life. For example, gears on the shafts of gears on the shaft of a concrete mixer, forging machine etc.

2.5 FIT OR LIMIT SYSTEM

A fit or limit system consists of a series of tolerance arranged to suit a specific range of sizes and functions, so that limits of size may be selected and given to the mating components to ensure specific classes of fit. It has two categories.

(*i*) **Hole basis system:** In this system, the limits on the hole are kept constant and the variations necessary to obtain the classes of fit are obtained by varying limits on the shafts.

(*ii*) **Shaft basis system:** In this system, the limits on the shaft are kept constant and the variations necessary to obtain the classes of fit are obtained only by varying the limits on the holes.

The holes are normally made by drilling, followed by reaming. Therefore, the dimension of a hole is fixed due to the nature of the tool used. On the other hand, shafts can be accurately produced to size by turning and grinding. Thus, the dimension of a shaft can be easily controlled by manufacturing processes. For this reason, the hole basis system is much more popular than the shaft basis system except where temperature may have a detrimental effect on large sizes.

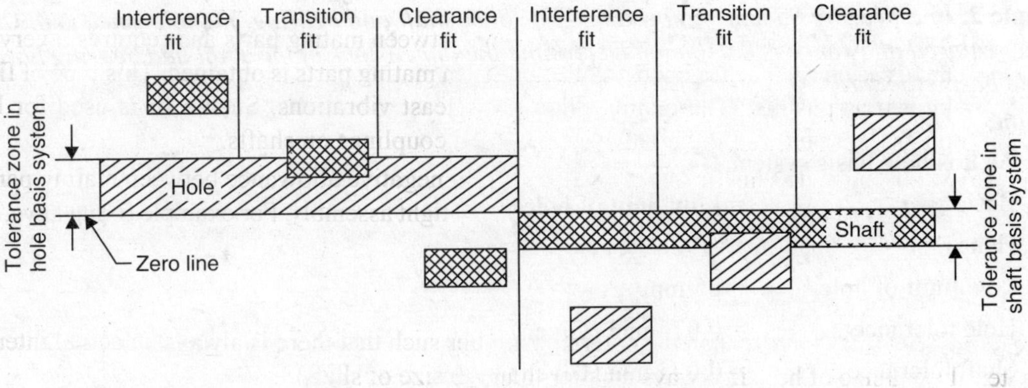

Figure 2.7 Shaft and Hole basis system

2.6 SOLVED EXAMPLES

Example 1. *Find the values of hole tolerance, shaft tolerance and allowance for the following dimensions of mating parts according to the hole basis system.*

Hole: 38.500 mm *Shaft: 38.470 mm*
38.523 mm *38.445 mm*

Solution:

High limit of hole = 38.523 mm
Low limit of hole = 38.500 mm
Hole tolerance = High limit of hole – Low limit of hole
= 38.523 – 38.500 = 0.023 mm
High limit of shaft = 38.470 mm
Low limit of shaft = 38.445 mm
Shaft Tolerance = High limit of shaft – Low limit of shaft
=38.470 – 38.445 = 0.025 mm

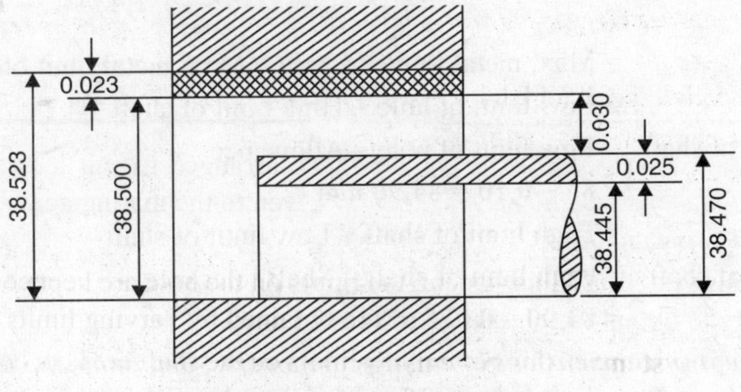

Figure 2.8 Solved Example (1)

Allowance = Max. metal limits of hole – Max. metal limit of shaft
= Low limit of hole – High limit of shaft
= 38.500 – 38.470 = 0.030 mm

Example 2. *In a bearing 85 mm shaft rotates. For both shaft and bearing, the tolerance is 0.075 mm and the required allowance is 0.010 mm. Determine the dimensions of the shaft and bearing bore with the hole basis standard.*

Solution:

As it is hole basis system,

Basic size = Low limit of hole

Hence,

Low limit of hole = 85 mm

Hole tolerance = 0.075 mm (Given)

Shaft tolerance = 0.075 mm (Given)

But,

Hole tolerance = High limit of hole – Low limit of hole

Or, High limit of hole = Low limit of hole + Hole tolerance

= 85 + 0.075 = 85.075 mm

Allowance = 0.10 mm (Given)

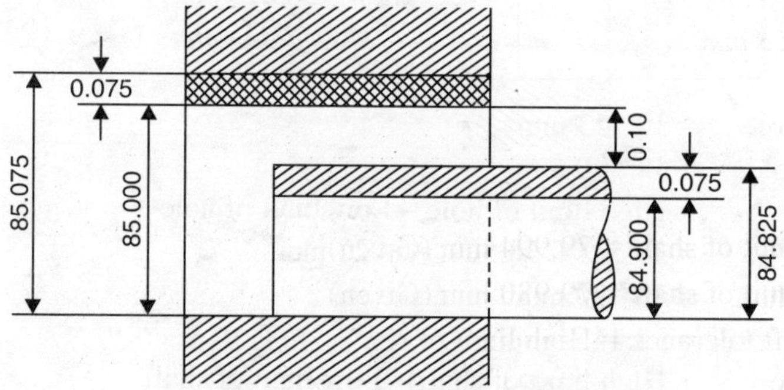

Figure 2.9 Solved Example (2)

But,

Allowance = Max. metal limits of hole – Max. metal limit of shaft

= Low limit of hole – High limit of shaft

Or, High limit of shaft = Low limit of hole – Allowance

= 85 – 0.10 = 84.90 mm

Shaft Tolerance = High limit of shaft – Low limit of shaft

Or, Low limit of shaft = High limit of shaft – Shaft tolerance

= 84.90 – 0.075 = 84. 825 mm

Example 3. *For the following dimensions of mating components, determine the following according to hole basis standard: basic size, hole tolerance, shaft tolerance, maximum clearance and minimum clearance.*

Hole:	*80.040 mm*	*Shaft:*	*79.994 mm*
	80.000 mm		*79.980 mm*

Solution:

According to hole basis system,

Basic size = Low limit of hole

Low limit of hole = 80.000 mm (Given)

Hence,

Basic size = 80.000 mm

High limit of hole = 80.040 mm (Given)

Therefore,

Hole tolerance = High limit of hole – Low limit of hole

= 80.040 – 80.040 = 0.040 mm

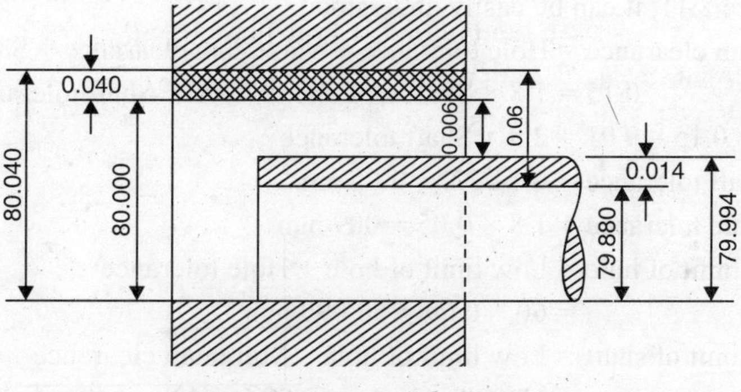

Figure 2.10 Solved Example (3)

High limit of shaft = 79.994 mm (Given)

Low limit of shaft = 79.980 mm (Given)

Shaft tolerance = High limit of shaft – Low limit of shaft

= 79.994 – 79.980 = 0.014 mm

Maximum clearance = High limit of hole – Low limit of shaft

= 80.040 – 79.980 = 0.06 mm

Minimum clearance = Low limit of hole – High limit of shaft

= 80.000 – 79.994 = 0.006 mm

Example 4. *The basic size of hole-shaft assembly is 60 mm, the maximum and minimum clearance are 0.15 mm and 0.01 mm respectively. If hole tolerance is 1.8 times shaft tolerance, find out the limits for hole and shaft accordingly to the hole basis standard.*

Solution:

According to hole basis system,

Basic size = Low limit of hole

Low limit of hole = 60.00 (Given)

Maximum clearance = 0.15 mm (Given)

Minimum clearance = 0.01 mm (Given)

Hole tolerance = 1.8 × Hole tolerance (Given)

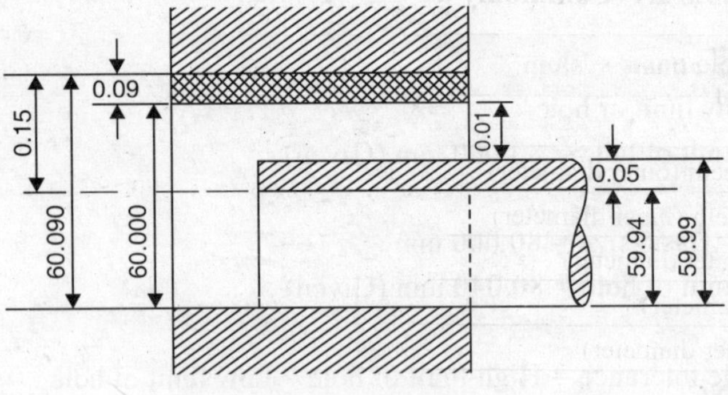

Figure 2.11 Solved Example (4)

From the Figure 2.11, it can be easily understood.

Maximum clearance = Hole tolerance + Maximum clearance + Shaft tolerance

or, 0.15 = 1.8 × Shaft tolerance + 0.01 + Shaft tolerance

or, 0.15 – 0.01 = 2.8 × Shaft tolerance

or, Shaft tolerance = 0.14/2.8 = 0.05 mm

Hence, Hole tolerance = 1.8 × 0.05 = 009 mm

High limit of hole = Low limit of hole + Hole tolerance

= 60 + 0.09 = 60.090 mm

High limit of shaft = Low limit of hole – Minimum clearance

= 60,000 – 0.01 = 59.99 mm

Low limit of shaft = High limit of hole – Maximum clearance

= 60.090 – 0.15 = 59.94 mm

2.7 ISO SYSTEM OF LIMITS AND FITS

This system allows for 27 types of fits and 18 grades of tolerances (IT 01, IT 0, IT1 to IT16) covering a size range of 0 to 3150 mm. In this system, 28 possible holes are designated by A, B, C, CD, D, E, EF, F, FG, H, J, JS, K, M, N, P, R, S, T, U, V, X, Y, Z, ZA, ZB and ZC while the shafts are designated by corresponding small letters. For specifying any particular hole or shaft, the rule is to write the letter followed by the number denoting the tolerance grade. For example, *F*6 is a hole of *F* type having a tolerance grade IT6. Similarly, *h*6 is a shaft of *h* type having a tolerance grade of IT6. A fit involving these two is written as *F*6-*h*6 or *F*6/*h*6.

Either a hole basis or shaft basis system can be obtained from this system. It is recommended that a hole basis system should be used while a shaft basis system is used only in some cases.

2.8 COMMONLY USED HOLES AND SHAFTS

In actual practice, the fits required can be met by a quite small selection from the full range available in the standards. Commonly used, holes and shafts are as follows:

Holes: *H*6, *H*7, *H*8, *H*9, *H*11

Shafts: *c*11, *d*10, *e*9, *f*7, *g*6, *h*6, *k*6, *n*6, *p*6, *s*6.

Table 2.1 Commonly used fits for various applications

Application	*Type of Fit*
1. GEAR	
With external teeth (outer diameter)	*h*8
With internal teeth (inner diameter)	*H*9
Worm wheel (outer diameter)	*h*8
Worm (outer diameter)	*h*8
Worm rack (inner diameter)	*H*9
2. BUSH BEARING	
On outer diameter	*H*7/*h*6
On inner diameter	*H*7/*js*6
3. BALL/ROLLER BEARING	
Gear Transmission	
Housing	*H*7
Shaft	*k*6 up to 40*d*
	*k*6, *js*6 from 40 to 100*d*
	*m*6 from 100 to 250*d*
Centrifugal Pump	
Housing	*J*7
Shaft	*k*6 up to 40*d*
	*k*6, *js*6 from 40 to 100*d*
	*m*6 from 100 to 250*d*
Belt Conveyor	
Housing	*M*7, *K*7
Shaft	*g*6
Crane Wheel	
Housing	*N*7
Shaft	*n*6, *m*6
Spindle of Grinding Machine	
Housing	*K*7, *J*7
Shaft	*k*6, *js*6 upto 40*d*
	*k*6, *js*6 (ball bearing) from 40 to 100*d*
	*m*6 (roller bearing) from 40 to 100*d*
	*n*6 over 100*d*
4. LEAD SCREW AND NUT	
Lead screw (outer diameter)	*h*8
Nut (inner diameter)	*H*9
5. DOVEL PINS	*H*7/*js*6

SUMMARY

- The term interchangeable manufacture means that the parts which go into assembly may be selected at random from large number of parts. These parts have their dimensions within certain specified limits.
- The advantage of interchangeable manufacture is that the components will assemble together at random.
- Shaft refers not only to the diameter of circular shaft but also to any external dimension of a component.
- Hole refers not only to the diameter of a circular hole but also to any internal dimension of a component.
- Basic size is the size which is obtained by calculation from design consideration.
- Actual size is the dimension as measured on a manufactured part (shaft or hole).
- The maximum size permitted for the part is known as maximum limit or high limit.
- The minimum size permitted for the part is known as minimum limit or low limit.
- Tolerance is the difference between maximum and minimum limits of the component, *i.e.* between high limit and low limit.
- Allowance is an intentional difference between the maximum material limit of mating part. For shaft, the maximum material limit will be its high limit and for hole, it's low limit.
- In unilateral system, both the limits of size are on the same side of the zero line.
- In bilateral system, one of the limits of size is on one side of the zero line and the other limit of size is on the other side of the zero line.
- In clearance fit, size of hole is always larger than the size of shaft.
- In transition fit, tolerance zone of shaft and hole intersect.
- In interference fit, size of hole is always smaller than the size of shaft).
- In hole basis system, the limits on the hole are kept constant and the variations necessary to obtain the classes of fit are obtained by varying limits on the shafts.
- In shaft basis system, the limits on the shaft are kept constant and the variations necessary to obtain the classes of fit are obtained only by varying the limits on the holes.

ASSIGNMENT

1. Students should be shown different types of fits used in different applications.

EXERCISES

1. What do you understand by interchangeable manufacture?
2. With help of conventional diagram of limits and fits, define the following:

   (*i*) Tolerance (*ii*) Allowance
   (*iii*) Basic size (*iv*) Zero line
   (*v*) Tolerance zone.
3. Define the following terms:

   (*i*) Actual deviation (*ii*) Fundamental deviation
   (*iii*) Basic shaft (*iv*) Basic hole.

4. Illustrate the clearance, interference and transition fit.
5. Differentiate between hole-basis and shaft-basis system of fits.
6. Give sub-classification of clearance fit with examples.
7. What do you mean by interference fit? Also, give its classification.
8. Explain the difference between allowance and clearance.
9. Differentiate between upper and lower deviation.
10. Explain also unilateral and bilateral tolerance. State their advantages and disadvantages.
11. In a limit system, the following limits are specified to give a clearance fit between a shaft and a hole.

Hole:	30.000 mm	Shaft:	29.995 mm
	30.002 mm		29.982 mm

Determine the following according to hole basis standard: basic size, shaft tolerance, hole tolerance, maximum clearance and minimum clearance.

[**Ans.** Basic size 30 mm, Shaft tolerance 0.013 mm, Hole tolerance 0.020 mm, Maximum clearance 0.038 mm and Minimum clearance 0.005 mm]

12. A fit on 65 mm shaft requires a hole tolerance and shaft tolerance each equal to 0.225 mm and an interference of 0.0375 mm. Determine the proper hole and shaft dimensions according to the hole basis system.

[**Ans.** L.L. of hole 65 mm, H.L. of hole 65.225 mm, L.L. of shaft 65.2625 mm and H.L. of shaft 65.4875 mm]

❑❑❑

Installation of Machine Tools

3

CHAPTER

3.1 INTRODUCTION

Modern industry needs a large number of machine tools to produce various items. The quality of product depends upon the accuracy of machine tool used (apart from the skills of the operator). To achieve accuracy and to produce part economically, various tests are performed on machine tools as follows:

(*i*) Alignment or Geometrical tests deal with manufacturing accuracies.

(*ii*) Performance or Practical tests deal with the working accuracies.

Collectively, these tests are known as acceptance tests. These tests are applied on new and reconditioned machines. Correct installation is the first essential step before testing a machine tool for various acceptance tests. Installation of a machine tool consists of checking up the machine for completeness. It involves operations as under:

(*i*) Reading of information manual supplied by the manufacturer of machine.

(*ii*) Suitable location to be decided.

(*iii*) Erection drawing or proper foundation plan is to be laid down.

(*iv*) Preparing the foundation followed by its curing.

(*v*) Slinging of the machine tool.

(*vi*) Lifting and erecting the machine tool on the foundation.

(*vii*) Levelling of the machine tool.

(*viii*) Connecting the foundation bolts and grouting them properly.

(*ix*) Second levelling after the foundation bolts are set.

(*x*) Final checking of level before testing.

3.2 READING OF INFORMATION MANUAL

Information manual are supplied by the manufacturer of the machine to be installed. These manual should be read carefully. Information manual contains important information which includes specifications of machine, drawings, installation details, various test and maintenance jobs. This information is very important for correct installation of machine.

3.3 LOCATION

Location is one of the most important step of installation. Location of a machine tool means selection of a suitable place, area etc., where the machine tool can be installed. The location of a machine tool in a shop depends upon the availability of area for the installation and the required working area around the machine tool without interferring with surrounding machines or the workers around it. Location should be dry, clean and well ventilated. The space available should be sufficient enough for the unrestricted working of the operator, free movement of all machine parts taking into consideration the maximum traverse of the table and saddle of the machine tool along with an adequate storing space for the raw material and finished parts.

3.4 FOUNDATION OF MACHINE TOOLS

Foundation is an important part of a structure which transmits the weight of that structure to the ground. There is need for a better foundation for machine to ensure improved performance. All structures constructed on land are well supported on a foundation, a foundation, is thus, a connecting link between the structure and ground. Hence, it should to be made very carefully.

Foundation of machine tool performs the following functions:

(*i*) It serves as the base to support the entire load of the machine tool, workpiece, tools accessories, mountings etc.

(*ii*) It absorbs mechanical vibrations produced during the operation.

(*iii*) It prevents transmission vibrations from one machine tool to another.

(*iv*) It helps to maintain correct alignment of machine tool, its mechanism and moving parts.

(*v*) It provides such height to the machine tool which makes its operation convenient to the operator.

General requirements of machine foundations are as listed below:

(*i*) The foundation should be able to carry the superimposed loads without causing shear or crushing failure.

(*ii*) No resonance should occur; hence the natural frequency of foundation soil system should be either too large or too small compared to the operating frequency of the machine tool.

(*iii*) Machine foundation should be separated from the adjacent machine's foundations.

3.5 DIFFERENT TYPES OF MACHINE FOUNDATIONS

Four types of machine foundations are commonly used (Figure 3.1), such as:

3.5.1 Block Type

It is a solid foundation of cement concrete or brick masonry, in which, foundation bolts are grouted and the machine rest over the top surface. The mass of this type of foundation is large. Such types of foundations are used for machines producing impulsive and periodical forces at low speeds.

3.5.2 Box Type

This provides a hollow box type construction usually of cement concrete. The machine is placed over the top surface. The mass of this type of foundation is less as compared to block type.

3.5.3 Wall Type

It may be constructed of cement concrete of brick masonry with a base slab and pair of walls over which there is a top slab. Machine rests on the top slab.

3.5.4 Frame Type

With a fabricated frame at the top, it is similar to wall type foundation. The machine is placed and supported over the frame work at the top. This type of foundation is most suitable for machine having rotating motion and high speed.

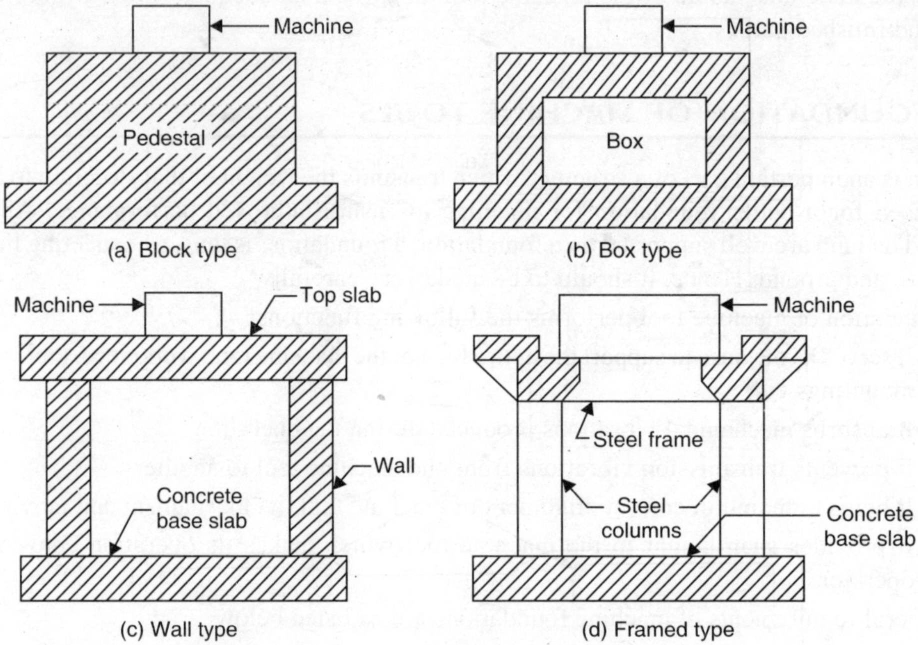

Figure 3.1 Different types of machine foundations

3.6 FACTORS AFFECTING THE TYPE AND SIZE OF FOUNDATION

Separate foundations are not required for different machine tools, if the strong concrete floor is available. These machines can be directly installed on the floor only. In machine tools, in which excessive mechanical vibrations are produced a separate foundation of either concrete or brick is required. The choice of a particular type of foundation is determined by the different factors like weight, type of soil, power source, vibrations produced, direction and nature of motion etc.

(*i*) **Weight of the machine.** Heavy machine needs strong foundation.

(*ii*) **Type of soil.** The load bearing capacity of soils is different.

(*iii*) **Power source.** Deeper foundation of machine tool is needed for a higher power driving motor.

(*iv*) **Mechanical vibrations produced.** Vibrations in the machine tools are caused by various factors and can be very detrimental not only for machine and operator, but also for precision work. Heavy and stronger foundations are required for high vibration.

(*v*) **Direction and nature of motion.** A horizontal machine tool needs heavier foundation as compared to vertical machine tool.

3.7 FOUNDATION PLAN (ERECTION DRAWING)

While designing the machine foundations, we must have full information about machines, *i.e.*, nature, size, weight and other characteristics. Foundation plan shows the profile and size of the necessary concrete preparation and gives particulars as to recesses required to accommodate projections and bolts etc. It is also necessary to know the local conditions such as space available, soil condition and about other supporting structure available. The use of such a drawing enables a template to be made so that the foundation can be dug and set down before the machine is delivered and will be ready for operation as soon as it arrives.

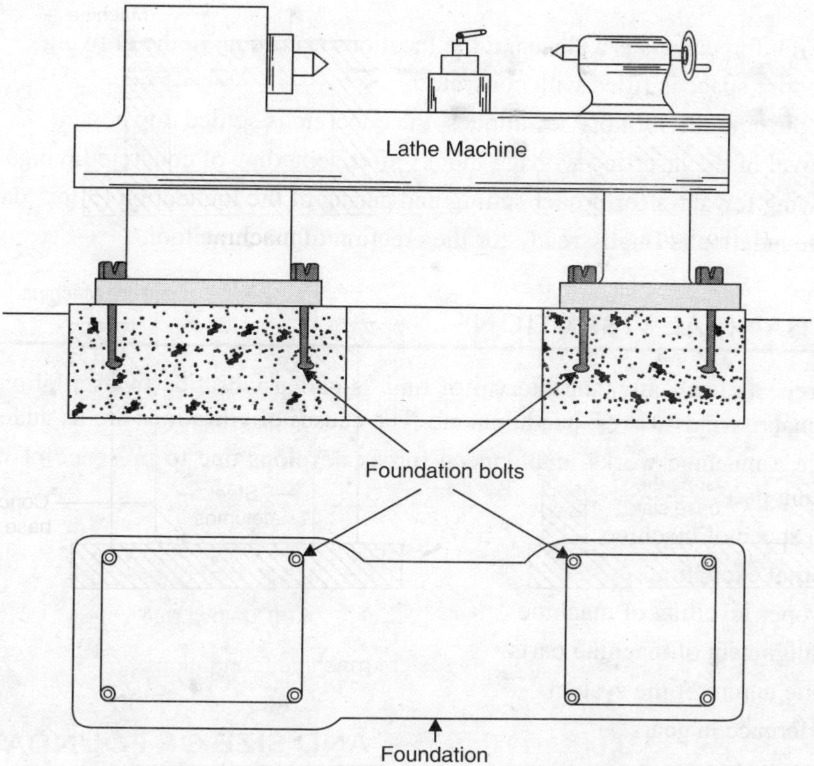

Figure 3.2 Foundation Plan for Lathe Machine

Figure 3.2 shows foundation plan for a light duty lathe machine. Spate block is used for head stock end and tail stock end.

Foundation plan is a detailed drawing, which consists of following information related to the machine foundation:

(*i*) Size and profile of the machine tool.
(*ii*) Weight of the machine.
(*iii*) Shape and size of the foundation to be prepared.
(*iv*) Type of foundation to be prepared and the material required (RCC or bricks).
(*v*) Location of mechanism, area needed for opening of base doors, operators' position etc.
(*vi*) Location of holes and recesses to accommodate the projections and foundation bolts in machine base, if any.
(*vii*) Location of electric motor, power supply, cables etc.
(*viii*) Location of other accessories.

3.8 PREPARING THE FOUNDATION

After the foundation plan is ready the operations that are to be carried out for preparing the foundation are as follows:

(*i*) The length and width of the pit is marked.

(*ii*) The pit is dug out as per dimensions.

(*iii*) To level bottom of the machine tool, bottom pad is laid down in the pit.

(*vi*) Wooden blocks are placed at the locations where holes or projections or recesses are to be made.

(*v*) Steel reinforcements are placed at the locations according to the drawing.

(*vi*) The entire space is filled with concrete.

(*vii*) With the help of vibratory techniques, the concrete is settled down.

(*viii*) Removal of the inserting wooden blocks after sometime of concrete laying.

(*ix*) Allowing few days for proper setting and curing of the foundation (3 to 5 days).

(*x*) The foundation is finally ready for the erection of machine tool.

3.9 MECHANICAL VIBRATIONS

A motion that repeats itself after an interval of time is called vibration or oscillation like impact of hammer, earthquake, winds waves, pendulum etc. The causes of vibrations are as under:

(*i*) While a machine works, unbalanced forces develops due to presence of reciprocating or rotating part.

(*ii*) High speed of machine.

(*iii*) External excitation.

(*iv*) Improper levelling of machine.

(*v*) Misalignment of machine parts.

(*vi*) Elastic nature of the system.

(*vii*) Interference in gears.

(*viii*) Hydraulic forces.

(*ix*) Self-induced vibrations due to cutting process.

(*x*) Loose foundation bolts of foundation or loose parts of the machine.

Each cause of vibration generates a force which is changing in either its direction or magnitude. It is the force which causes vibrations and resulting vibrations characteristics will be determined by the manner in which forces are generated.

3.9.1 Types of Vibrations

There are two types of vibrations—free and forced vibration.

(*i*) **Free vibration or natural vibration:** When no external source acts on the system, after giving it in an initial displacement, it will vibrate. These vibrations are known as free or natural vibrations. Frequency of these vibrations is called natural frequency of vibration. Sources of natural vibration are wind, waves, earthquake, blow of hammer etc.

(*ii*) **Forced vibration:** These vibrations are caused due to external excitation. Source of forced vibration: Engines, air compressor, printing press, generating set etc.

3.9.2 Effects of Vibrations

The adverse effects of vibrations are as follows:

(*i*) Unbalanced forces of machine, causing vibrations, can also damage the machine parts and can cause accidents.

(*ii*) Vibrations may cause harmful effects on the adjoining machines.

(*iii*) The vibrations transmitted to the building can damage it, if the frequency of vibration is close to the natural vibrations of the building.

(*iv*) The vibrations cause rapid wear of machine parts such as bearing, gears etc.

(*v*) Unwanted vibrations can also cause loosening of machine parts.

(*vi*) Vibrations are harmful for machine operator.

(*vii*) The proper reading of instruments cannot be taken due to vibrations.

(*viii*) Vibrations cause annoyance to the person in an area around the machine.

Remark

(*i*) If the frequency of excitation coincides with any one of natural frequencies, condition of resonance is reached, during which large vibrations may occur which can result in failure of the system.

(*ii*) Critical speed (RPM) of the machine is the speed at which the vibration coincides with the natural frequency of the machine. There may be more than one critical speed for rotary components.

(*iii*) Beneficial effects and uses of vibrations.

(*a*) In various manufacturing processes like casting, forging, and welding processes.

(*b*) In settling down concrete cement mixture.

(*c*) One of the most important uses of vibrations is found in musical instruments.

3.10 DAMPING OF VIBRATIONS

Damping means resistance to the motion of a vibratory system. The following methods may be used to damp out of vibrations in machine tools.

1. Alignment of machine tool should be perfect, as misalignment causes vibrations.
2. The foundation bolts should be kept tight and the machine should be bolted rigidly to the foundation block.
3. Slab of resilient material (such as rubber, cork) should be provided between machine and foundation or foundation and ground. The resilient material provides damping effect. Following two methods are possible:

(*i*) In the first method, the slab of resilient material is placed between the machine tool and foundation as shown in Figure 3.3. The resilient material may be cork, linoleum, rubber or any other substance.

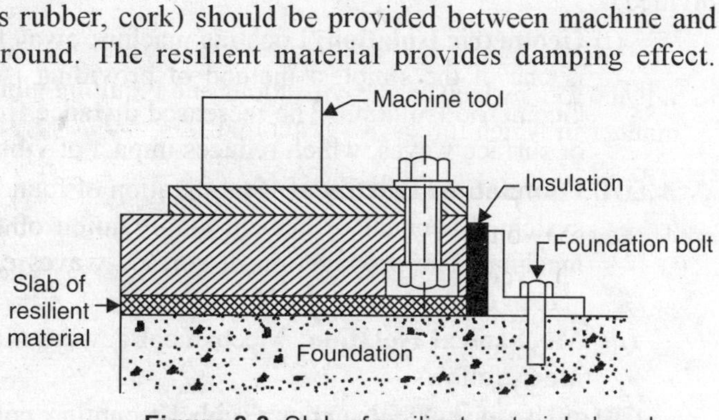

Figure 3.3 Slab of resilient material placed between the machine tool and its foundation

(*ii*) In the second method, the resilient material is placed between the foundation and ground so as to place the machine tool directly over the foundation.

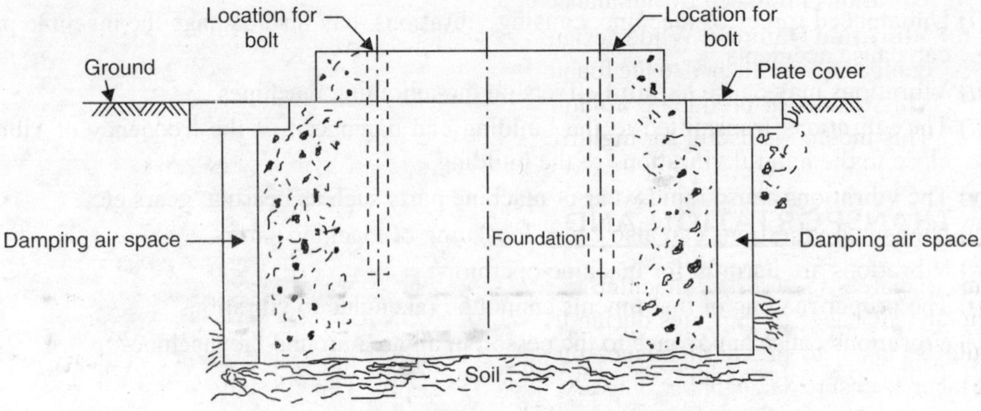

Figure 3.4 Foundation block with damping air spaces

4. Foundation block can be isolated by providing air spaces, which provide cushioning effect. Air contained in the damping air spaces functions as a damping medium. Figure 3.4 shows foundation of this type. Some part of vibrations generated can be absorbed by these air spaces. In order to provide further damping, foundation block can be made to rest over resilient material.

3.11 VIBRATION ISOLATION

Vibrations produced during operation of a machine tool should be reduced to below the permissible level of vibrations to keep the machine tool in good conditions, to increase the productivity and to achieve better precision on the workpiece surface.

The transmission of vibrations can be controlled by isolating either the sources (producing vibration) or the receiver.

Isolation of source is known as active isolation and the protection of receiver is known as passive isolation. When the machine is to be prevented from transmitting the vibration, active isolation is used. While the passive isolation is a sensitive machine itself is to be protective from vibrations from external sources like other machine or traffic etc.

There are three categories into which vibration isolation methods for machine tools may be divided:

(*i*) **Geometric Isolation:** Locating machine away from adjoining structures and establishments is one of the simplest method of providing isolation. This type of isolation is known as Geometric Isolation. The increased distance from the machine helps in reducing amplitude of surface waves, which reduces impact of vibration.

(*ii*) **Foundation Treatment:** Stabilization of foundation soil increases natural frequency of the system. Chemical or cement stabilization of sandy soil is useful for reducing frequency machine. The propagation of surface waves can be obstructed significantly by providing screen or sheet piles.

(*iii*) **Mechanical Isolation:** Mechanical devices may also be used to control vibrations, which are as follows:

(*a*) **Use of shock absorber:** Rubber mounting, cork layer, plastic pads, steel springs are directly placed between the machine and the foundation or the foundation and the base, as absorbers which help in reducing vibrations.

(*b*) **Counter balancing of loads:** In many reciprocating machines, it is possible to counter balance the exciting load (unbalanced load) by providing counter load. This help in reducing vibrations produced by unbalanced load.

(*c*) **Vibration Damper:** While designing the foundation additional masses (known as vibration dampers) are attached to the foundation. This makes system a multi-degree freedom system. This results in producing a natural frequency much different from operating frequency. This method is useful for high frequency machines.

3.12 TRANSPORTATION AND ERECTION

At the time of purchase of a new machine tool, its initial inspection is carried out to find out if there have been any shortages or damages during transportation from supplier side to installed side. Proper care should be taken to prevent damages during transportations. All necessary corrective measures should be taken to ensure damage-free delivery. When the machine tool is ready for erection, the important job is to transport it safely to the foundation (erection site).

There are two possibilities.

(*i*) The packed machine tool is received through transportation, and then, directly unloads at the site.

(*ii*) The packed machine tool is first received and unloaded stored and then transported to the site.

In first case, the important point is to unload the machine tool and place it directly over the prepared foundation. Thus, firstly, the machine tool is unpacked and lifted. The common equipments used for loading and unloading of machine consist of different types of pulley block: like the rope pulley block, western pulley block and worm block pulley (shown in Figure 3.5).

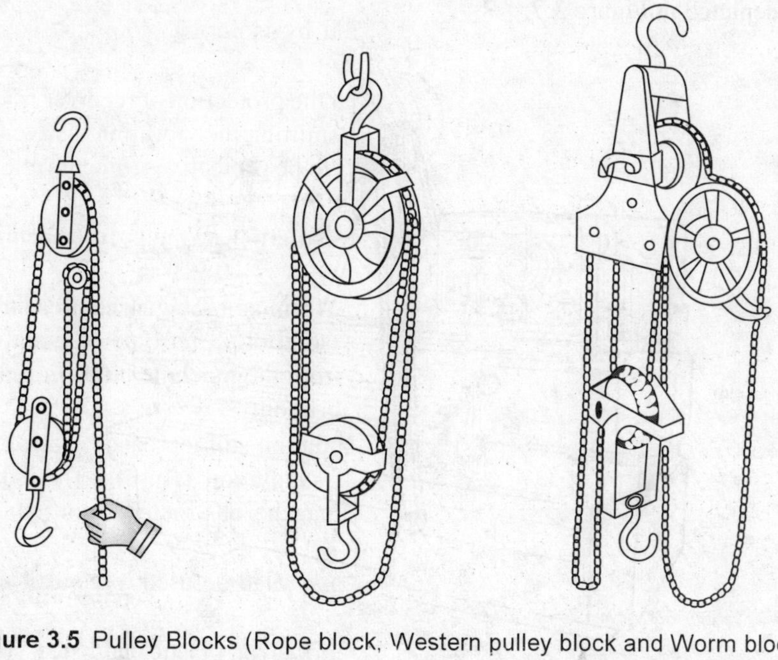

Figure 3.5 Pulley Blocks (Rope block, Western pulley block and Worm block)

Selection of the most suitable pulley block depends on the weight of machine tool.

(*i*) **Rope Pulley Block:** They are used for low loads (upto 1 ton).

(*ii*) **Western Pulley Block:** They are used low and medium loads (upto 1-5 tons).

(*iii*) **Worm Block Pulley:** They are used for heavy loads (upto 5 tons).

The pulley block must be grouted properly in a wall either supported at both end or may be of cantilever type. For heavier loads, cranes are used. Care should be taken to ensure that the weight to be lifted is within the capacity of the lifting device used.

In the second case, the machine tool is to be transported to the erection site after re-lifting from the source with the help of mobile crane. It can be reloaded in another truck and transported internally to the site. Other means of transportations like hand-trolley, lift truck, conveyors are used, in accordance with the size and width of the machine tool.

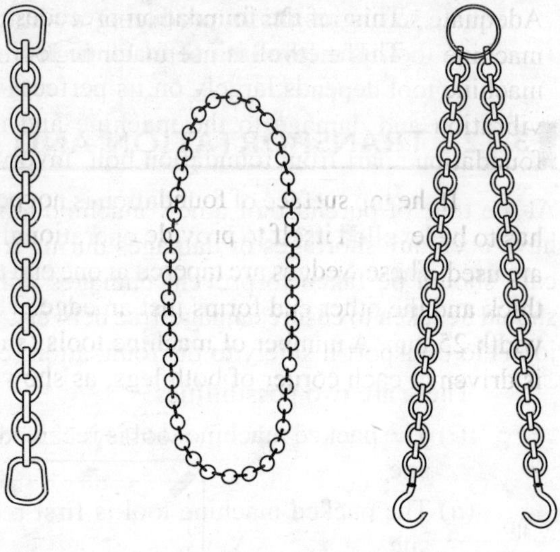

Figure 3.6 Chain Slings (Collar sling, Endless sling, Double sling)

Thus, in both cases, before lifting the machine tool, its proper slinging is necessary. Assortments of slings are required for this. These slings can be made up of rope, wire or chain and are made in various lengths from 1.5 m upwards. The common types of chain slings are collar sling, endless sling, double sling (shown in Figure 3.6). Each sling has a definite load carrying capacity and care should be taken to note this before using. The proper way of slinging lathe machine is depicted in Figure 3.7.

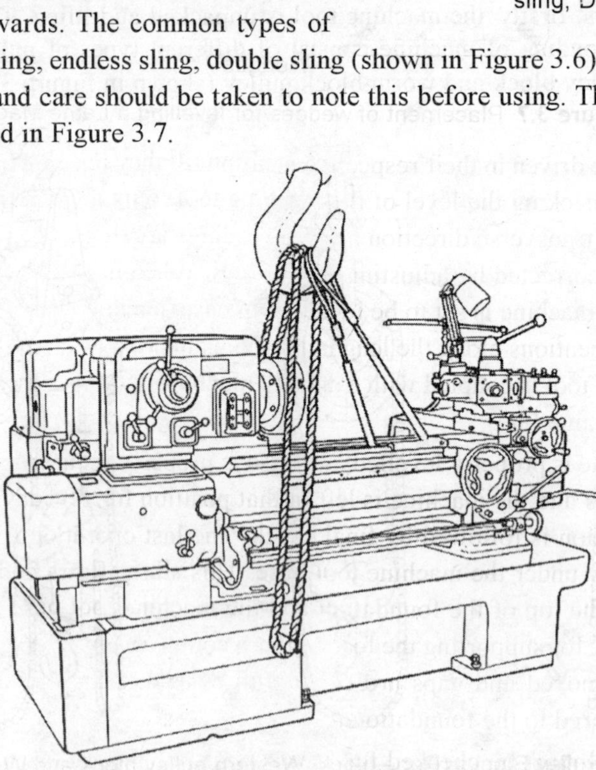

Figure 3.7 Proper way of slinging a Lathe Machine

Satisfactory completion of the erection work requires prior planning of the movements involved, a good team-work and professional experience.

3.13 LEVELLING AND ALIGNING

Adequate setting of the foundation precedes placing of machine in position, leading to levelling of the machine tool. The levelling of machine tool is to be done very carefully because the accuracy of the machine tool depends largely on its perfect levelling. Any error in levelling and alignment may cause vibration and damage to the machine and the foundation. The installation of machine tool on the foundation apart from foundation bolt, involves the use of levelling pad, shoes and wedges.

If the top surface of foundation is not perfectly levelled, the machine tool placed on the foundation has to be levelled itself to provide operational accuracy. To level, the machine a number of steel wedges are used. These wedges are tapered at one end and flat at another, such as the thicker end is 3 mm to 5 mm thick and the other end forms just an edge. The length of wedges may vary from 50 mm to 80 mm and width 25 mm. A number of machine tools (such as the Lathes) are supporting on two legs. One wedge is driven at each corner of both legs, as shown in Figure 3.7.

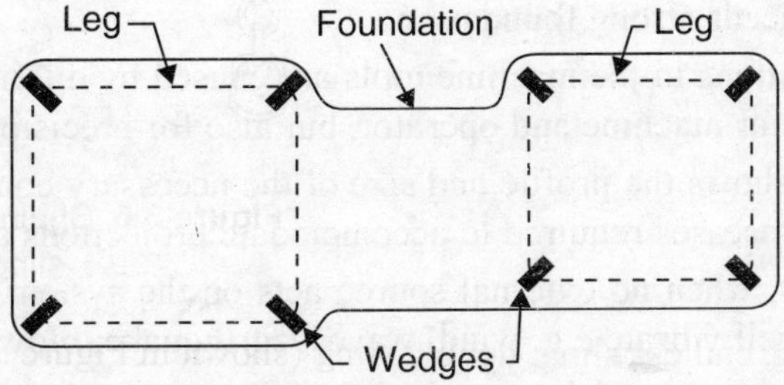

Figure 3.7 Placement of wedges for levelling a Lathe Machine

These wedges are driven in their respective position till they become tight under its base. A good spirit level is used for checking the level of the machine tool. This level is placed on the machine bed both in longitudinal and transverse direction and the readings are noted. Deviations from zero reading, if any, are immediately corrected by adjusting the wedges. When in both the directions the spirit level shows zero reading, the machine is set to be levelled. In case, machine has a very long bed, spirit level is placed at number of locations along the length as well as in cross-section to ensure perfect levelling. Sometimes, the machine tool is coupled with a driving machine and in such case, the centre line of both must be brought to the same line.

When the machine is properly levelled, the base is properly secured by tightening the loose nuts over the foundation bolts and the machine is left in that position for few days to allow, the machine to settle down. This operation is followed by final check. The last operation is grouting, which involves pouring of cement slurry under the machine tool base. This slurry flows and fills up all the voids and empty spaces between the top of the foundation and the machine tool base. It thus, provides a larger solid area under the base for supporting the load. After a couple of days, the grouting will become hard, then the wedges are removed and gaps are filled with cement. Thus, the machine tools is perfectly levelled and tightly secured to the foundation.

The machine tool must be checked for levelling and alignment after about every six month to take care of any misalignment due to operation.

SUMMARY

- To achieve accuracy and to produce part economically, acceptance tests are performed.
- Correct installation is the first essential step before testing a machine tool for various acceptance tests.
- Before installing machine, its information manual should be read carefully. This manual contains information regarding specifications of machine, drawings, installation details, different tests and maintenance jobs.
- The location of a machine should be sufficient enough for the unrestricted working of the operator and free movement of all machine parts with an adequate storing space for the raw material and final products.
- Foundation is an important part of a structure which transmits the weight of that structure to the ground.
- Foundation serves as the base to support the entire load of the machine tool, workpiece, tools accessories, mountings etc. It absorbs mechanical vibrations produced during the operation. It also prevents transmission vibrations from one machine tool to another.
- Heavy machine needs strong foundation.
- Mechanical vibrations in the machine tools are caused by different factors and can be very harmful not only for machine and operator, but also for precision work.
- Foundation plan shows the profile and size of the necessary concrete preparation and gives particulars as to recesses required to accommodate projections and bolts etc.
- In free vibrations, when no external source acts on the system, after giving it in an initial displacement, it will vibrate *e.g.* wind, waves, earthquake, blow of hammer etc.
- Forced vibrations are caused due to external excitation *e.g.* engines, air compressor, printing press, generating set etc.
- If the frequency of excitation coincides with any one of natural frequencies, condition of resonance is reached, during which large vibrations may occur which can result in failure of the system.
- Alignment of machine tool should be perfect, as misalignment causes vibrations.
- The foundation bolts should be kept tight and the machine should be bolted rigidly to the foundation block.
- In order to damp out vibrations, slab of resilient material should be provided between machine and foundation or foundation and ground.
- Isolation of source is known as active isolation and the protection of receiver is known as passive isolation.
- The machine tool must be checked for leveling and alignment after about every six month to take care of any misalignment due to operation.

ASSIGNMENT

1. Students should be taken to industries and shown various types of machine foundations.

EXERCISES

1. What operations are involved in the installation of machine tool?
2. Why reading of information manual of machine is important for correct installation?
3. What do you understand by the term location?
4. What are the functions performed by the foundation of machine tool?
5. With the help of neat sketches, describe different type of machine foundation.
6. What factors influence the size and type of a machine foundation and how?
7. What is the foundation plan? What information does it contain?
8. Describe the operation involved in preparation of machine tool foundation.
9. What do you understand by vibration? What are different causes of vibrations? What are harmful effects of mechanical vibrations?
10. What do you mean by damping of vibrations? What are different methods of damping vibrations of machine tool? Explain them.
11. What is different vibration isolation techniques used for machine tool? Explain each of them.
12. Write notes on transportation and erection of machine tools.
13. Describe the common type pulley block with neat sketch.
14. Why proper slinging of machine tool is necessary during erection?
15. Write notes on levelling and aligning of machine tools.

❑❑❑

Miscellaneous Examples of Installation

4

CHAPTER

4.1 INTRODUCTION

The manufacturer supplies with the equipment its information manual for the consumer. These manuals should be read very carefully. They contain important information regarding specifications of the equipment, its drawings, installation details, various tests and maintenance jobs. This information is extremely essential for correct installation of the equipment. The installation process has to be followed in accordance with the instructions given in the manual. This chapter describes and discussed some miscellaneous examples of installation.

4.2 INSTALLATION OF BELT DRIVE

A belt is one of the most common and effective device employed for transmission of motion and power from one shaft to another with the help of a thin inextensible band (belt) running over pulleys. A belt drive consists of three members- the driving pulley, the driven pulley and an endless belt (which envelops two pulleys). The pulley, mounted on driving shaft, is known as driving pulley or driver, while the one, which is mounted on the driven shaft, is called driven pulley or follower. In industrial machines, flat and V-shaped belts are commonly used. The flat belts are more versatile than the V-belts as they may be employed in various arrangements. The belt needs correct installation so as to get better performance.

The alignment and tension of belt should be carefully checked. Its regular inspection must also be carried out. The correct tension in the belt is a must in order to transmit the motion without slipping of belt. In case the belt is too tight, it will put excessive load on bearings. On the other hand, if the belt is too slack, it may result in easy slipping. The belt should run with the slack side on the top in order to provide increased arc of contact as shown in Figure 4.1.

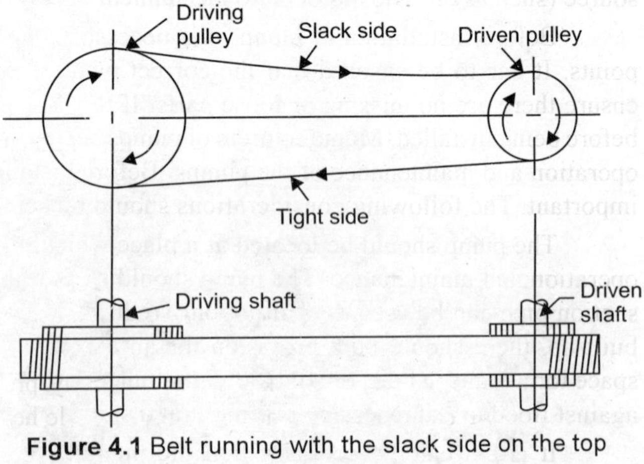

Figure 4.1 Belt running with the slack side on the top

It is necessary that the following checks should be made before starting a belt drive:

(*i*) The alignment of shafts should be checked carefully. Misalignment, if any, need to be immediately rectified for proper results.

(*ii*) The parallelism of the shafts should be checked by placing squares on the shafts and setting cord between the shafts along with the squares, as shown in Figure 4.2.

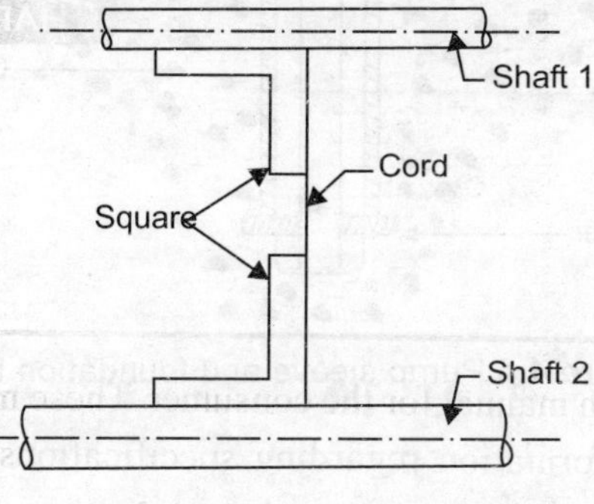

Figure 4.2 Check for parallelism of shafts

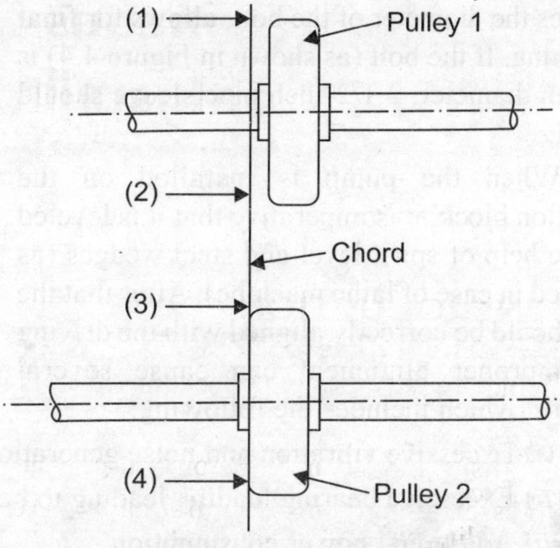

Figure 4.3 Alignment check for pulley

(*iii*) It is necessary to check the alignment of pulleys by using a cord along with their edges. If the pulleys are of the same width, the cord may contact with pulleys at four points (1), (2), (3) and (4) at each ends. Alignment check for pulleys is shown in Figure 4.3.

(*iv*) The pulleys should be given half-turn and checked again properly for alignment.

4.3 INSTALLATION OF PUMPS

A pump is a mechanical device that uses an external source of power to force fluid to move from one place to another. A pump develops no energy of its own, it only transforms energy from the external source (such as electric motor) into mechanical energy, which is manifested by the motion of the fluid.

Before installation of pump, it is necessary that one should scrutinize the following important points. It has to be ensured that the correct pump is delivered for the job. Next, check the pump to ensure there are no missing or loose parts. If the unit has a preservative casing, it should be removed before being installed. Manufacturers of pumps supply instruction manual with the pump for installation, operation and maintenance of the pumps. Before installing pumps, a thorough study of manual is very important. The following considerations should be kept in mind during installation of pump.

The pump should be located at a place which is easily accessible to the worker for installation, operation and maintenance. The pump should be placed near the liquid supply so that short and direct suction pipe can be used. It will not only reduce losses but also avoid formation of air pockets. In the built-up, there should be a provision for an ample head-room (a trap or ceiling opening) to provide space for a crane, a hoist or a tackle. Simultaneously, proper care should be taken to safeguard the pump against flood or dampness by placing it at a suitable height.

It is extremely necessary that the pump is leveled and secured to foundation properly. The foundation should have enough base area suitable with the bearing capacity of the soil. The foundation is usually made of concrete and extends beyond the pump base. It should be strong enough to absorb vibration and also to provide a rigid support to the pump and motor. Recommended mass of a concrete foundation should be three times that of the pump, motor and base. Foundation bolts are embedded in the concrete foundation. The bolts should always be longer than needed (3/4 to 1 inch) to allow for shimming up the pump to make it level and for grouting the base. A pipe sleeve which is about 2 1/2 times the diameter of the bolt, allows for final positioning. If the bolt (as shown in Figure 4.4) is 1 inch in diameter, 2 1/2-inch pipe sleeve should be used.

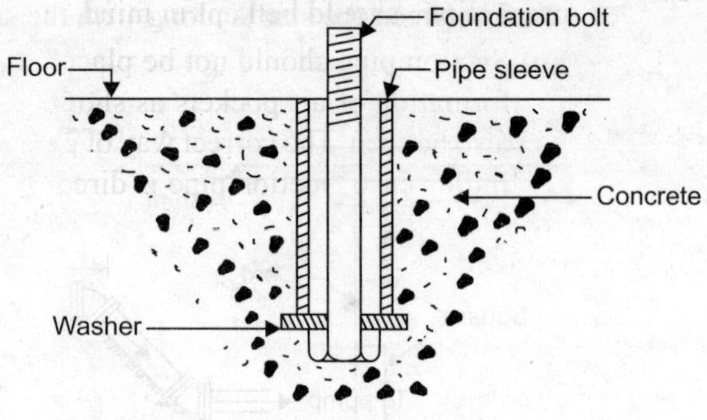

Figure 4.4 Pump sleeve and foundation bolt

When the pump is installed on the foundation block, it is imperative that it is leveled with the help of spirit level and steel wedges (as discussed in case of lathe machine). After that the pump should be correctly aligned with the driving unit. Improper alignment can cause several problems, which includes the following:

(*i*) Excessive vibration and noise generation.

(*ii*) Excessive bearing loading leading to bearing failure.

(*iii*) Additional power consumption.

(*iv*) Uneven wear and tear of mechanical components.

The coupling should be checked for angular and axial alignment. Angular misalignment is when the shafts are not parallel to one another while the axial dimension is the distance between the coupler flanges. The following instruments are needed for leveling and alignment work of the pump: feeler gauge, taper gauge, straight edge, spirit level and dial indicator.

Angular alignment is checked by inserting taper gauge between the coupling faces at each 90º around the faces as shown in Figure 4.5, whereas the axial alignment is checked by using feeler gauges with straight edges at each 90º rotation on the flange rims, as shown in Figure 4.6. Axial and angular alignment can also be checked by means of dial indicator and LASER. The piping connections should be done after tightening of the foundation bolts.

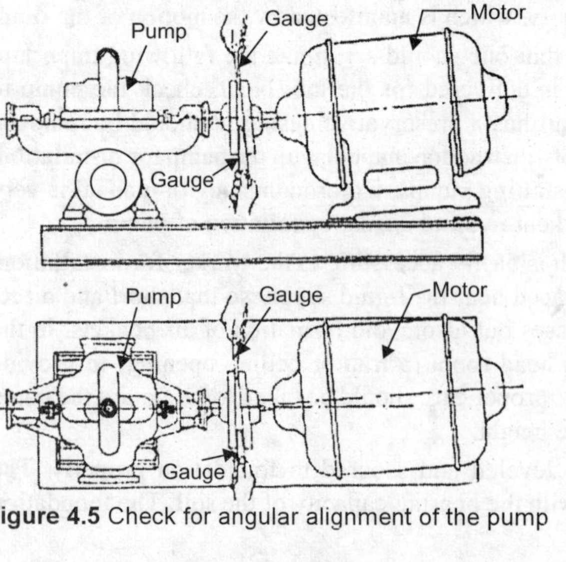

Figure 4.5 Check for angular alignment of the pump

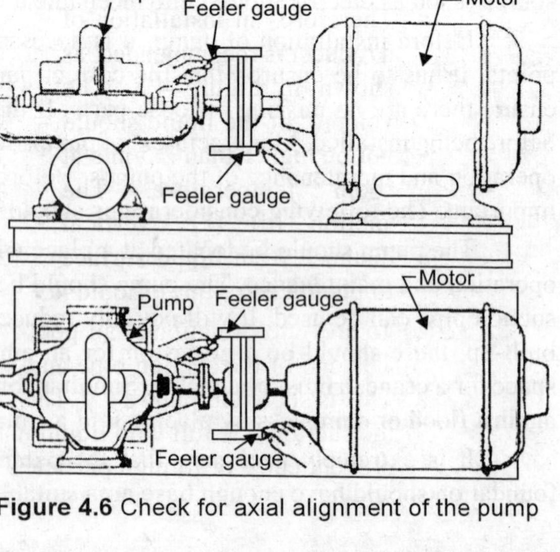

Figure 4.6 Check for axial alignment of the pump

4.3.1 Pipings used in Pumps

The three main types of pipings used in pumps as discussed below:

1. **Suction Piping:** The suction piping should be designed and connected in such a way as it always remains filled with liquid and no air pockets are formed inside it. The following precautions should be kept in mind.

   (*a*) Suction pipe should not be placed above the centre line of the pump as this may lead to formation of air pockets as shown in Figure 4.7(*a*). In this figure, joints and bends can also be seen. The correct way of placing the suction pipe is also shown in the Figure 4.7(*b*). In this case, suction pipe is direct without joints and bends.

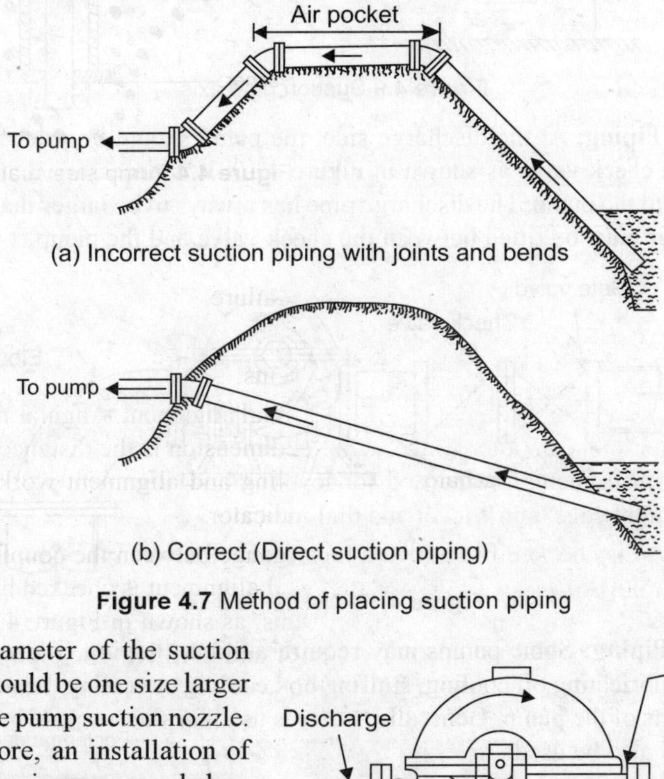

(a) Incorrect suction piping with joints and bends

(b) Correct (Direct suction piping)

Figure 4.7 Method of placing suction piping

   (*b*) The diameter of the suction pipe should be one size larger than the pump suction nozzle. Therefore, an installation of a reducer is recommended, as shown in Figure 4.8. The velocity of the liquid should not be higher than 2.5 m/s at room temperature. The suction lift (*i.e*. vertical distance from liquid surface to pump centre line) should not be above 6 m.

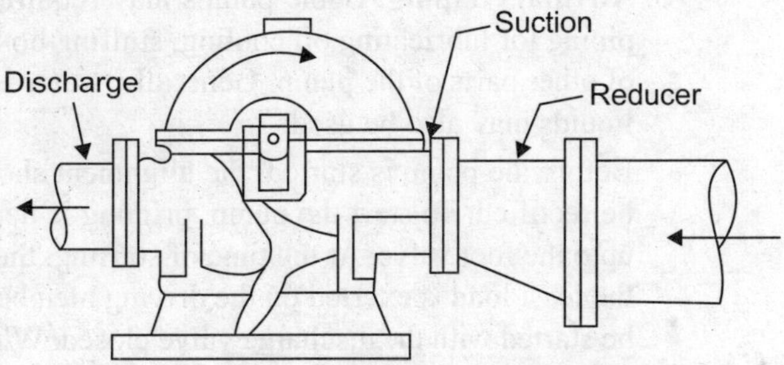

Figure 4.8 Suction pipe with a reducer

   (*c*) To prevent formation of air pockets, the slope of the suction pipe should be kept upward. The end of the suction pipe and the foot valve should be fully submerged in the liquid. This is shown in Figure 4.9. The check valve action of the foot valve is essential to keep the suction line full with liquid when the pump is idle. Suction strainers must be used to prevent entry of foreign particles into the pump.

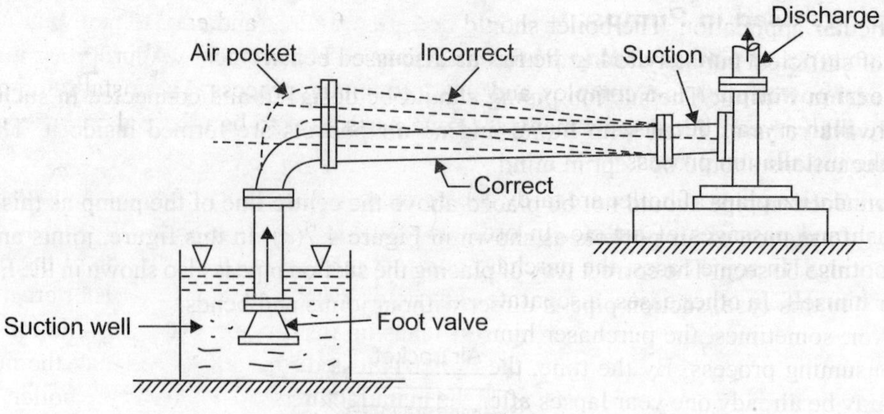

Figure 4.9 Suction pipe axis

2. **Discharge Piping:** At the discharge side, the pump should be protected by placing a gate valve and a check valve as shown in Figure 4.10. It is necessary that these valves must be fitted close to the pump. The discharge pipe has always to be larger than the pump discharge. An increaser must be fitted between the check valve and the pump.

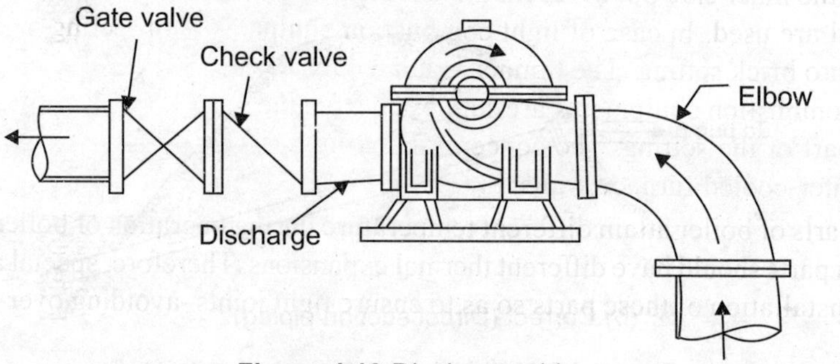

Figure 4.10 Discharge side

3. **Auxiliary Piping:** Some pumps may require auxiliary piping. These connections includes piping for lubricating oil cooling, stuffing box cooling or sealing, bearing cooling or cooling of other parts of the pump. Generally, water is used for this purpose but in some cases, other liquids may also be used.

   Before the pump is started, the alignment should be re-checked. Misalignment, if any, must be rectified. To start the pump, priming is necessary. The pump should be filled with liquid upto the foot valve. At the time of starting, the discharge valve is adjusted in such a way that the least load is exerted on the driving member. For radial type of impellers, the pumps must be started with the discharge valve closed. While for axial flow, mixed flow or propeller type of pumps, the discharge valve should be kept open. To avoid sudden variation of load on the driving member and to eliminate sudden surge in the discharge line, gradual closing or opening of a valve has to be done.

4.4 INSTALLATION OF STEAM GENERATOR (BOILER)

Steam generator, also known as boiler, is a unit in which steam is generated. It is a closed vessel made of steel or iron. The function of steam generator is to evaporate water at required condition by transfer of heat produced by burning of fuel. The boiler selected should be of adequate capacity to handle the load at a particular application. The boiler should be safe, efficient and easy to maintain. It is good to select boiler of sufficient capacity to handle future expansion requirements. Appropriate installation of boiler in a steam power plant is a complex and time consuming process. The installation process may consume more than a year, because its many separate parts have to be received, inventoried and then used during the installation process.

The foundation plans of boiler are tardy and complicated, as they require provisions for adequate disposal of ash and passages of air etc. In order to carry highly concentrated loads, it may require foundation footings. In some cases, the purchase contract may include installation of the boiler by the manufacturer himself. In other cases, a separate contractor, who is specialist in boiler erection work, is hired. However, sometimes, the purchaser himself takes up installation work. Installation as stated is very time-consuming process. By the time, the boiler is tested for compliance with the manufacturer guarantees, may be already one year lapses after the manufacturer had delivered the boiler. Boiler must be installed on a non-combustible floor. It should not be installed on carpet, wood or plastic. Floor should be preferably of concrete and properly designed to support the heavy load. The pipe and fittings on the steam line and blow-off lines should not be galvanized iron.

The term "setting" of the steam generator includes all the material required for surrounding the heating surfaces and the furnace. Usually, the setting is built up of a solid brick wall with the refractory material lining on the inner side but in recent times, refractory tiles on the inner side with layers of heat insulation material are used. In case of light combustion equipments, reflecting arches and other parts are usually built into brick setting. The thinner insulated material setting is employed for supporting it. However, heavy combustion equipments are supported from the foundation itself. Baffles, ash pits, soot hoppers etc. are part of the setting. The concept of setting has been modified to great extent with the introduction of water-cooled furnace walls.

Different parts of boiler attain different temperature during operation of boiler like tubes, baffles, tile walls etc. Such parts should have different thermal expansions. Therefore, special attention is required in designing and installation of these parts so as to ensure tight joints, avoiding over-straining tubes and cracking of walls.

The foundation of boiler is designed in accordance to the standard structural practice to support the load imposed. A properly erected steam generator will have its weight adequately dispersed and supported with adequate provision for thermal expansion. It is completely encased air-tight walls capable of preventing heat loss. Figure 4.11 depicts different types of furnace walls. The air space acts as insulation. The walls are pierced with holes at various locations for observation ports, gauge leads, soot blowers etc. Figure 4.12 shows Babcock and Wilcox boiler.

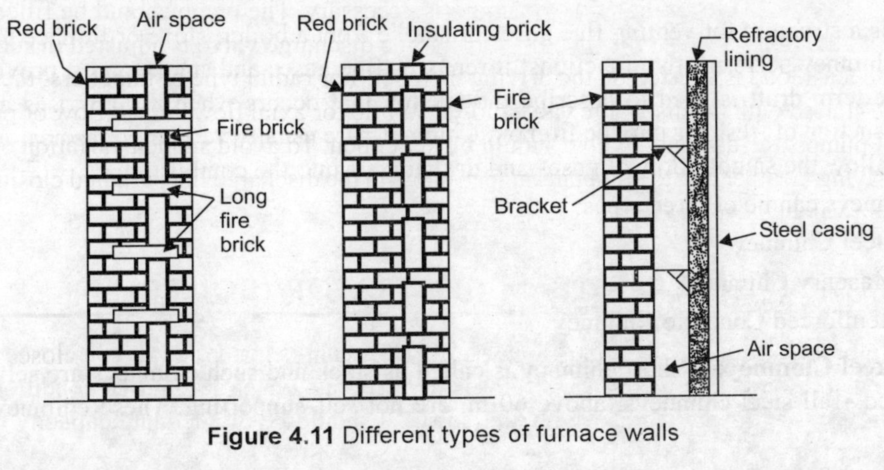

Figure 4.11 Different types of furnace walls

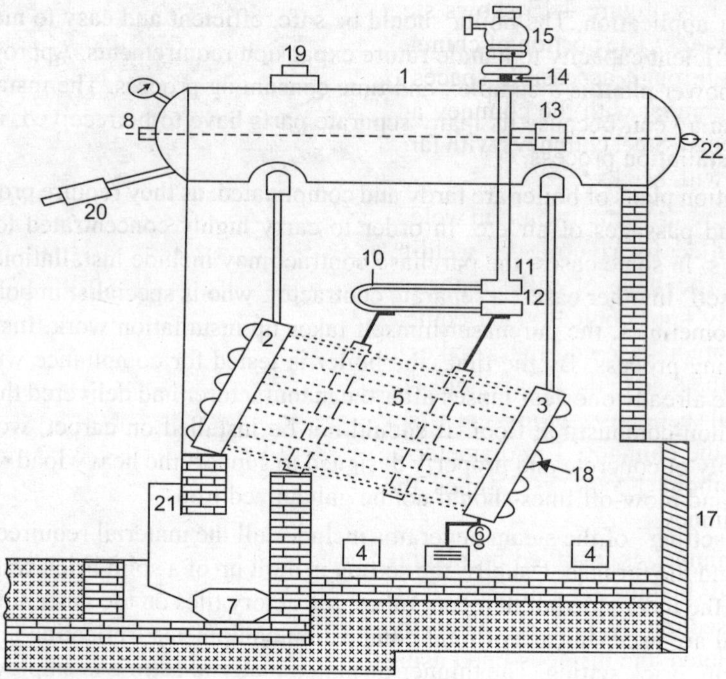

Components of Boiler:

1. Steam and water drum
2. Uptake header
3. Downtake header
4. Door
5. Tubes
6. Mud box
7. Fire grate
8. Water level indicator
9. Pressure gauge
10. Steel tubes
11. Superheated steam box
12. Saturated steam box
13. Dry pipe
14. Outlet pipe
15. Stop valve
16. Battles
17. Damper
18. Cap
19. Safety valve
20. Feed check value
21. Hopper
22. Chain

Figure 4.12 Babcock and Wilcox boiler

4.5 INSTALLATION OF CHIMNEY

A chimney is a structure for venting flue gases or smoke from a boiler, stove or furnace to be released outside. A chimney performs two functions: it removes flue gases and smoke and it provides draft for the fire. The term draft is used to describe the suction that occurs when hot air rises and creates a vacuum for suction of fresh air into the firebox. Chimneys are usually vertical, or as near as possible to vertical, to allow the smooth flow of gases and drawing air into the combustion.

Chimneys can be of three types:

(*i*) Steel Chimney

(*ii*) Masonry Chimney

(*iii*) Reinforced Concrete Chimney

(*i*) **Steel Chimney:** A short chimney is called as stack and such chimneys are self- supporting and un-guyed. Tall steel chimneys, above 60 m, are not self-supporting. These chimneys should be braced with guy wires. Figure 4.13 shows steel chimney with guy wires. The portion of chimney which may extend through accessible spaces is enclosed to avoid contact with the chimney and damage to the chimney. Steel chimneys with large diameter are lined with bricks.

Steel chimneys are usually built in welded sections and installed by riveting or welding horizontal joints. The base of self-supporting chimneys is usually kept conical. This conical shape provides a larger base and thus, makes it simpler to provide the load to the foundation. The height of the conical section is usually kept 1/6 of the total height of the chimney. The diameter at the base is kept about 1.5 to 1.75 times the diameter of the chimney.

The steel chimney is installed by placing a floating scaffold inside the chimney. This scaffold is raised section by section. When the steel work is completed, the inside is lined using scaffold. After completion of the installation work, it is painted with two coats of steel paint. It should be repainted after every two years for required services. The steel chimneys are lined with suitable refractory material to enhance their functional benefit. The lower section is made with fire bricks and fire clay while the upper section is lined with common brick and cement mortar. The lining is supported on steel angles riveted to the inside of the chimney.

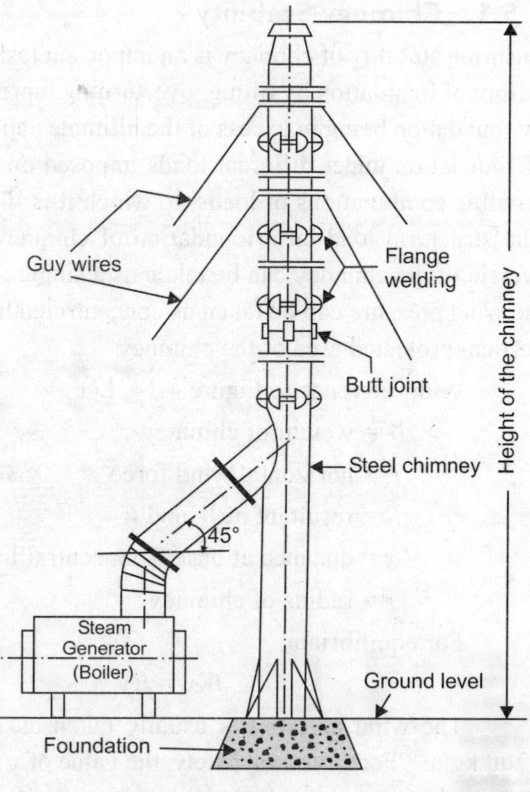

Figure 4.13 Steel Chimney with guy wires

The width of plates used in steel chimneys is about 1.25 to 2.25 m. For vertical seams, single riveted lap joints are used. For horizontal seams, single, double or triple riveted lap joints are used. However, for horizontal seams (with heavy plates) and the flare of the chimney, butt joints are used.

(*ii*) **Masonry Chimney:** Earlier, common bricks were used for masonry chimneys, but now-a-days, radial bricks are used. These bricks can be molded in accordance with the diameter of the chimney. Masonry chimneys are constructed of solid masonry, water-proof concrete block or rubble stone with mortar joints. Fire clay flue linings are employed in masonry chimneys. The thickness for brick and concrete block is 4 inch while for rubble stone is 12 inch. The section joints are bedded in refractory mortar. An air space of 1 inch is used to separate the flue liner from the masonry providing enough mortar to be used to make a good joint and hold the tiles in proper position.

(*iii*) **Reinforced Concrete Chimney:** The construction cost of reinforced concrete (RCC) chimney is much higher than masonry chimney because of the need for forming the steel structure. Also, concrete is not as impervious to extreme weather in comparison to hard burned bricks used in masonry chimneys. However, RCC chimneys are much lighter as compared to masonry chimneys. Architectural treatment can be given to masonry and reinforced concrete chimneys for the beautification purpose. It will look nice with the plant and its surroundings. This, however, is not possible in case of steel chimneys.

4.5.1 Chimney Stability

Ensuring stability of chimney is an important task. Stability analysis aims at removing the possibility of failure of foundation by tilting, overturning, uprooting and sliding due to load intensity imposed on soil by foundation being in excess of the ultimate capacity of the soil. It is essential to check for the stability of foundation under different loads imposed on it. The foundation should remain stable under all the possible combinations of loads, to which it is likely to be subjected under the most severe conditions. The structural load on a foundation of chimney is a result of chimney's weight and wind pressure. Weight of the chimney can be taken as a single vertical force acting through its centre of gravity while the wind pressure can be taken as concentrated horizontal force, acting through the centre of gravity of vertical projected area of the chimney.

With reference to Figure 4.14, Let

W = weight of chimney

F = horizontal wind force

R = resultant of W and F

x = distance at base from central line of chimney to the point where R cuts

r = radius of chimney

For equilibrium,

$$W.x = F.y$$

The wind pressure is usually taken as 100 kg/m^2 and radial brick masonry work weight as 2200 kg/m^3. For structural safety, the value of x must be less than r in case of masonry chimneys, but it may safely exceed r in other types of chimneys. Steel and RCC chimneys are stable even in the presence of high wind pressure since they have capability to withstand tension at the windward side of the shell B. But, masonry chimneys are not capable of withstanding any tension at this side. Therefore, for such chimneys, the resultant of chimney weight and wind force must lie within radius of the chimney. This is because, if distance x is greater than chimney radius r, then to prevent tipping of the chimney about side A due to wind force, there must be a pull at side B in downward direction. This pull is provided by the tension in the steel shell in steel chimneys and by the steel reinforcements in RCC chimneys. But, in case of masonry chimneys, such tensile forces are totally absent and hence, distance x must lie within radius of the chimney.

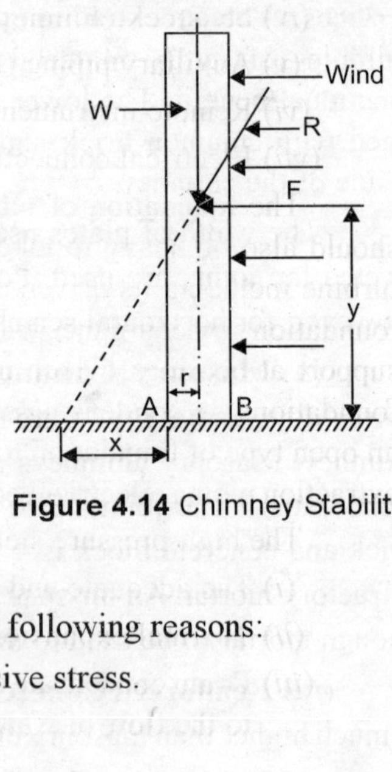

Figure 4.14 Chimney Stability

4.5.2 Causes of Chimney Failure

Great damage is done if chimneys fail. It may fail due to any one of the following reasons:

(*i*) Crushing at side A may take place due to excessive compressive stress.

(*ii*) Due to wind force, downward tipping may result.

(*iii*) In rare cases, horizontal sliding may take place.

(*iv*) Joints failure due to ageing or poor mortar may occur.

(*v*) Failure due to uneven settlement of foundation.

4.5.3 Foundation of Chimney

Figure 4.15 shows foundation of a chimney. Foundation of chimney should be of monolithic concrete having sufficient base area suitable with the bearing capacity of the soil. The shape of foundation is octagonal. It is kept larger than the chimney at the base top.

The width at the base must be sufficient to be consistent with the bearing capacity of the soil but should be less than 1/10 (chimney height + chimney diameter at the top). Depth of the foundation is generally made 4% of the chimney height subject to a minimum of 1.2 m.

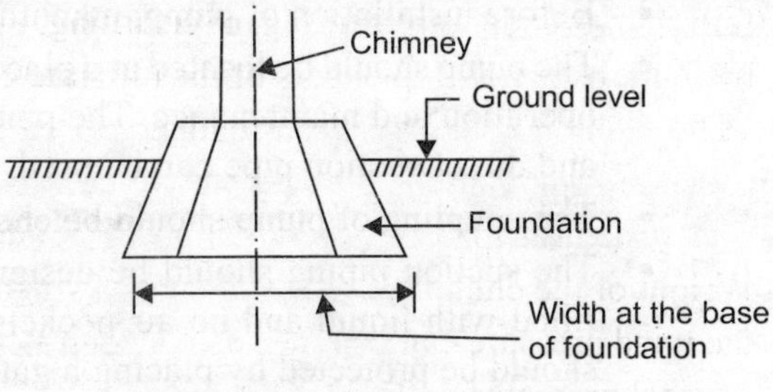

Figure 4.15 Chimney Foundation

4.6 INSTALLATION OF STEAM TURBINE

A steam turbine is a prime mover in which the potential energy of steam is transformed into kinetic energy, and after that kinetic energy transformed into mechanical energy of rotation of turbine shaft. The turbine shaft, directly or with the help of reduction gear is connected with the driven mechanism. The process of installation of a steam turbine includes the following steps:

(*i*) Laying out foundation.
(*ii*) Connection of steam from high pressure header.
(*iii*) Connection for exhaust.
(*iv*) Steam extraction piping.
(*v*) Auxiliary piping (for drip, drain, oil etc.)
(*vi*) Remote instruments and control wires.
(*vii*) Electrical connections for the generator and exciter.

The foundation of turbine should be rigid and capable of providing support to the turbine. It should also be heavy to take unbalanced forces produced during turbine operation. The small steam turbine including its driven equipment should be mounted on a rigid bed plate, which is erected on a foundation in accordance with manufacturer's design. The large steam turbines require fore and aft support at bearings, which may be in the form of brackets or pedestals and are carried on prepared foundations. In condensing steam turbines, exhaust usually flows downwards and hence, they require an open type of foundation to provide space just below the turbine, for items such as condenser, steam extraction piping, electrical connections etc. A steam separator should be installed just before the turbine.

The high-pressure steam pipe has to be installed with proper care to provide for:

(*i*) The adequate and flexible support.
(*ii*) Thermal expansion with it.
(*iii*) Drain condensate (especially essential during warm-up period to prevent turbine damage due to the flow of water along with the steam).

SUMMARY

- A belt is employed for transmission of motion from driving shaft to driven shaft with the help of a thin inextensible band (known as belt) running over pulleys.
- In industrial machines, flat and V-shaped belts are commonly used.

- The alignment and tension of belt should be correct checked. Shafts and pulleys must be properly aligned.
- Before installation of pump, manufacturer's instruction manual must be thoroughly studied.
- The pump should be located at a place which is easily accessible to the worker for installation, operation and maintenance. The pump should be placed near the liquid supply so that short and direct suction pipe can be used.
- The coupling of pump should be checked for angular and axial alignment.
- The suction piping should be designed and connected in such a way as it always remains filled with liquid and no air pockets are formed inside it. At the discharge side, the pump should be protected by placing a gate valve and a check valve.
- The boiler selected should be of adequate capacity to handle the load at a particular application. The boiler should be safe, efficient and easy to maintain.
- The foundation of boiler is designed in accordance to the standard structural practice to support the load imposed.
- Boiler must be installed on a non-combustible floor. Floor should be preferably of concrete and properly designed to support the heavy load. The pipe and fittings on the steam line and blow-off lines should not be galvanized iron.
- Different parts of boiler attain different temperature during operation. Such parts should have different thermal expansions.
- A chimney not only removes flue gases and smoke but also provides draft for the fire.
- Short chimneys are self- supporting and un-guyed. Tall steel chimneys, above 60 m, are not self-supporting. These chimneys should be braced with guy wires.
- Stability analysis of chimney is done with an aim of removing the possibility of foundation failure.
- The foundation of turbine should be rigid and capable of providing support to the turbine. It should be able to take unbalanced forces produced during turbine operation.

ASSIGNMENT

1. Students should be taken to industries and shown different types of installations as discussed in the unit.

EXERCISES

1. State the important aspects of installation of the belt drive.
2. State the sequence of operations to be followed in installation of pump.
3. Discuss about foundation of pump.
4. With the help of neat sketches, explain how angular and axial alignment of pump is checked.
5. Discuss various pipe connection of a pump.
6. Write notes on installation of steam generator.
7. What do you understand by chimney? Explain different types of chimneys.
8. What is meant by chimney stability? Explain the concept chimney stability in case of different types of chimneys.
9. What are the causes of chimney failure?
10. Write brief notes on chimney foundation.
11. Explain how a steam turbine is installed.

Machine Tool Testing

5 CHAPTER

5.1 INTRODUCTION

After installation of machine tool, testing is the next operation before its commissioning. The machine tools are subjected to following two tests, collectively called as acceptance tests:

(*i*) Alignment or Geometrical test which deals with manufacturing accuracies.

(*ii*) Performance or Practical test which deals with the working accuracies.

These tests are conducted on both new and reconditioned machines. The manufacturer usually provides a manual along with the machine tool, containing information regarding the foundation plan, guidance for installation and the test charts.

The test charts provide a sort of guarantee that the machine tool has been tested in such conditions under which it is required to work. The test charts display the details of necessary tests required to be conducted at work place and the results obtained thereof. Permissible alignment and working accuracy attainable are clearly specified in these charts.

The present day test procedure is based on the test chart prepared by Dr. G. Schlesinger in 1927. These charts specify the test procedure for different machine tools. Acceptance tests of machine tools include the following:

(*i*) No load and load test.

(*ii*) Tests for operations of different mechanisms.

(*iii*) Tests for rigidity of the machine tool.

(*iv*) Tests for productive output of the machine tool.

(*v*) Tests for alignment accuracy of the machine tool.

(*vi*) Tests for accuracy and surface finish of the product being machined.

5.2 SITES FOR TESTING

To confirm results of the chart, the tests given in the test chart are repeated at the user's site, provided that necessary testing equipments and skilled workmen are available. The most appropriate site for testing is the manufacturing site itself, as maximum care is taken there with respect to the adjustment and alignment of the machine tool during its assembly. Besides this, another advantage of testing machine tools at the manufacturing site is the availability of skilled workmen and important equipments required for testing.

5.3 MACHINE TOOL TESTS

Machining of a given geometric surface is achieved primarily by the combination of work-tool movements. In simple words, we can say that the accuracy of the machined surface depends upon the accuracy of the mating elements present in the machine tool. The different tests conducted on any machine tool shall include:

(*i*) Tests for checking of installation of machine in horizontal or vertical directions.

(*ii*) Tests for flatness of machine bed and straightness and parallelism of bed ways or bearing surfaces.

(*iii*) Tests for perpendicularity of guideways with respect to other guideways or bearing surfaces.

(*iv*) Tests for the true running of the main spindle and its axial movements.

(*v*) Tests for parallelism of spindle axis with respect to guideways or bearing surfaces.

(*vi*) Tests for the linear movement of different members, *e.g.*, saddle and table cross-slides etc., along their ways.

(*vii*) Practical tests for checking accuracy and surface finish of the product being machined.

Dial indicator, tests mandrels, spirit levels, straight edges and squares, autocollimators, alignment telescope, water level, wavinessmeter etc. are the instruments needed to conduct these tests.

5.4 MEASURING INSTRUMENTS USED FOR ALIGNMENT TEST

The most important condition for selection of any measuring instrument is its capability to measure with the required degree of accuracy under specified conditions. The measuring instruments generally used for alignment test are briefly described below:

5.4.1 Dial Indicator

Dial indicator is shown in Figure 5.1. It is also called as dial gauge or indicating gauge. It is widely used for alignment tests to check parallelism and concentricity of holes or rods and flatness of surfaces to an accuracy of 0.01 mm. Dial indicator should be mounted on strong base to ensure proper stability. The magnetic-base stand is the generally used. Dial indicator is graduated into 100 divisions. A full revolution of the indicator plunger represents 1 mm travel of the plunger. Thus, 1 division of the dial represents 0.01 mm travel of the plunger.

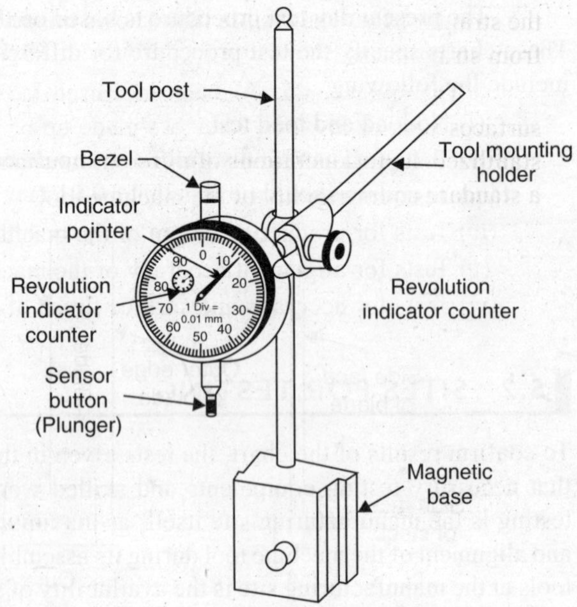

Figure 5.1 Dial Indicator

5.4.2 Test Mandrel

To check the true running of the spindle, test mandrels are used. They should be heat treated, stress relieved and made to a length which varies from 100 to 600 mm. They should be accurate and effect of natural sag should not be overlooked. The mandrel to be used must be as light as possible and must be straight and cylindrical to start with. To reduce the weight, hollow test mandrels may be employed. Two types of test mandrels are used:

(*i*) Mandrels with a cylindrical measuring surface and a taper shank which can be inserted into the taper bore of the main spindle.

(*ii*) Cylindrical mandrels which can be held between centers.

5.4.3 Spirit Level

Spirit level as shown in Figure 5.2 is used for levelling the machine and is based on precision angle measuring principle. It consists of sealed glass tube ground on its inside surface enclosed in a wooden or metal body. The tube is partially filled with spirit such that a bubble of air is left inside. This bubble comes to the top when spirit level is placed on surface. A centre is marked at the centre of glass tube. When the surface is flat and levelled, the bubble comes exactly in the centre of mark. These are available in various sizes and specified by the length, *e.g.*, 200 mm spirit level.

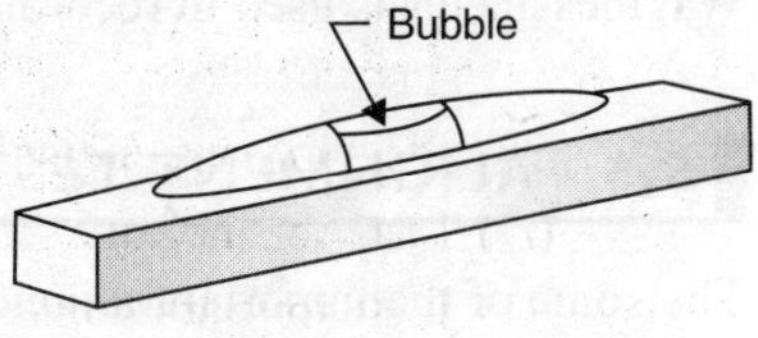

Figure 5.2 Spirit Level

5.4.4 Straight Edge and Square

The straight edge is like a solid bow. It is placed on machined surface to check them for flatness or straightness. Straight edges are made up of cast iron or steel. They should be heavy, well ribbed and free of internal stresses. Also, they should have wide bearing surfaces and machined straight flat edges.

The accuracy in straightness and flatness is tested by placing the straight edge on the surface to be tested and watching the light between the surface and the straight edge. The visible light shows the deviation from straightness.

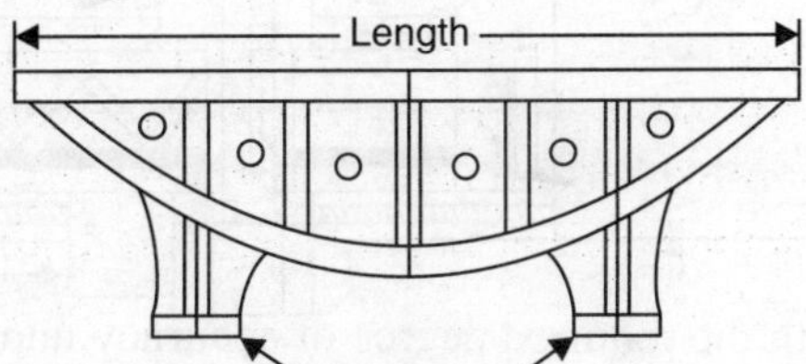

Figure 5.3 Straight Edge

The square is a precision tool used for laying out lines or for testing the squareness of two surfaces with each other. They are made up of good quality tool or alloy steel, suitably heat treated, stabilized, lapped and finally ground. It should be free from cracks, flaws or pits. The error at the top of a standard square should be less that +0.01 mm, of a precision square less than +0.005 mm.

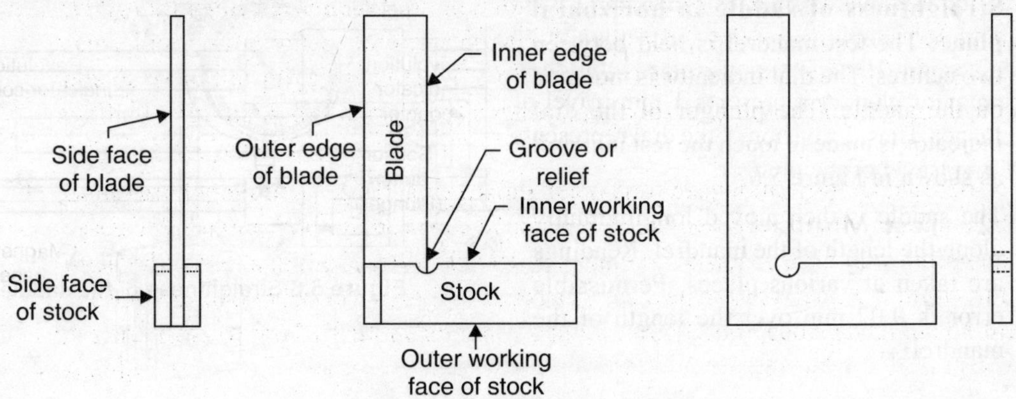

Figure 5.4 Engineer's Square

5.4.5 Autocollimator

Autocollimator is used for checking deflections of long beds in horizontal, vertical or inclined planes.

5.4.6 Alignment Telescope

Optical alignment telescope is used to indicate errors of alignment in both the vertical and horizontal planes of the optical axis.

5.4.7 Wavinessmeter

Wavinessmeter is used in recording and examining the surface waviness.

5.5 ALIGNMENT TESTS ON LATHE

The some of the important alignment tests performed on the lathe machine are as follows:

1. **Level of the Machine:** Before conducting test on any machine tool, it is necessary that it should be installed correctly in horizontal and vertical planes on properly constructed foundation. It is essential that it must be isolated from the surrounding by using vibration dampers.

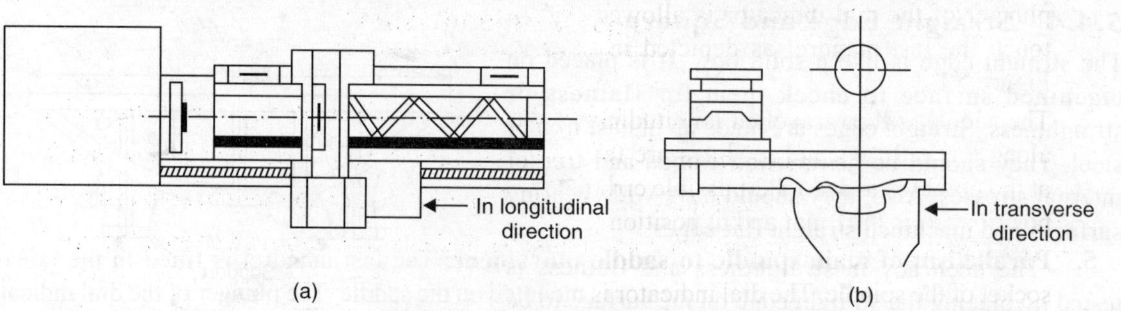

Figure 5.5 Level of installation

The level of the machine bed in longitudinal and transverse directions is checked by a spirit level. The spirit level is placed over the gauge block on the bedways at the front position, back position and in the crosswise direction. The position of the bubble in the spirit level is checked and the readings are noted. The setup for testing the level of a lathe machine is shown in Figure 5.5. Permissible error for front guideways is 0.02 mm/m convex while for rear guideways, it is 0.01 to 0.02 mm/m convex.

2. **Straightness of saddle in horizontal plane:** The test mandrel is held between two centres. The dial indicator is mounted on the saddle. The plunger of the dial indicator is made to touch the test mandrel as shown in Figure 5.6.

   The saddle is then moved longitudinally along the length of the mandrel. Readings are taken at various places. Permissible error is 0.02 mm over the length of the mandrel.

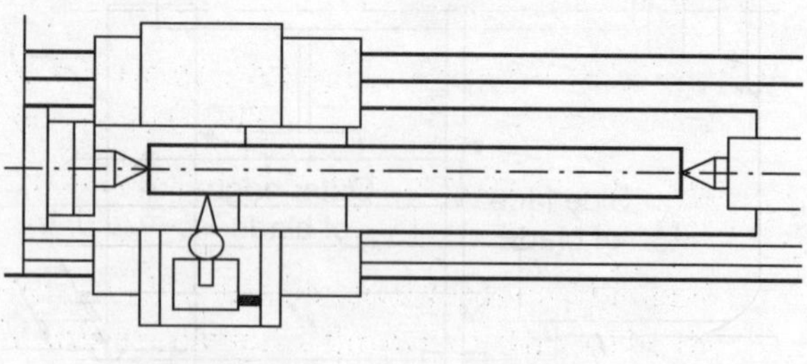

Figure 5.6 Straightness of the saddle

3. **Alignment of both centres in the vertical plane:** This test is carried out in vertical place. The test mandrel is fitted between two centres.

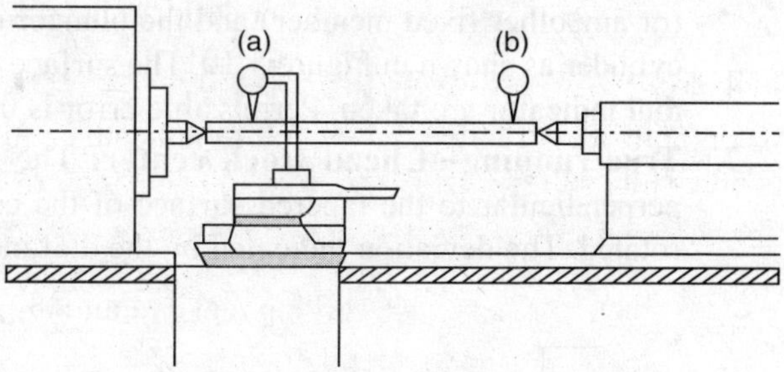

Figure 5.7 Alignment of both centers in the vertical plane

The dial indicator is mounted on the saddle in vertical place. The plunger of the dial indicator is allowed to touch the test mandrel as shown in Figure 5.7. The saddle along with the dial indicator is travelled longitudinally along the bedways over the length of the mandrel. The readings are taken at various places. Permissible error is 0.02 mm over the length of the mandrel.

4. **True running of taper socket in main spindle:** The test mandrel is fitted into tapered socket of the main spindle. The dial indicator is mounted in the saddle. The plunger of the dial indicator is allowed to touch the test mandrel as depicted in the Figure 5.8.

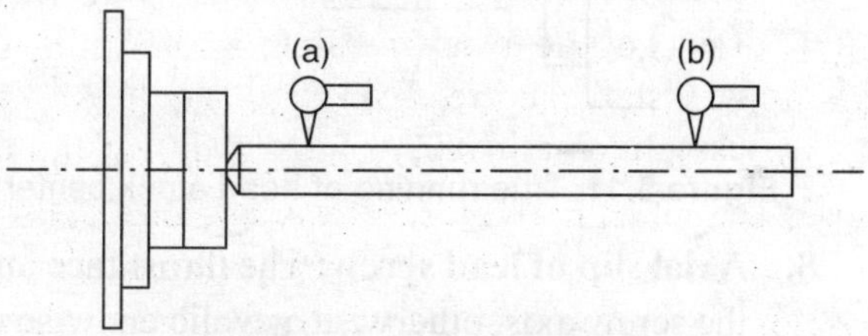

Figure 5.8 True running of taper socket in main spindle

The saddle is then travelled longitudinally along the bedways and readings are taken at the point (*a*) and (*b*). Permissible error at position (*a*) is 0.01 mm and at position (*b*), it is 0.02 mm.

5. **Parallelism of main spindle to saddle movement:** The test mandrel is fitted in the tapered socket of the spindle. The dial indicator is mounted on the saddle. The plunger of the dial indicator is allowed to touch the test mandrel and the saddle is moved to and fro.

It is checked in vertical as well as in horizontal plane. The indication in horizontal plane is given by (*b*) and in vertical plane by (*a*). Permissible error is 0.02 mm per length of test mandrel rising towards free end only and 0.02 mm per length of test mandrel inclined at free end towards tool pressure only.

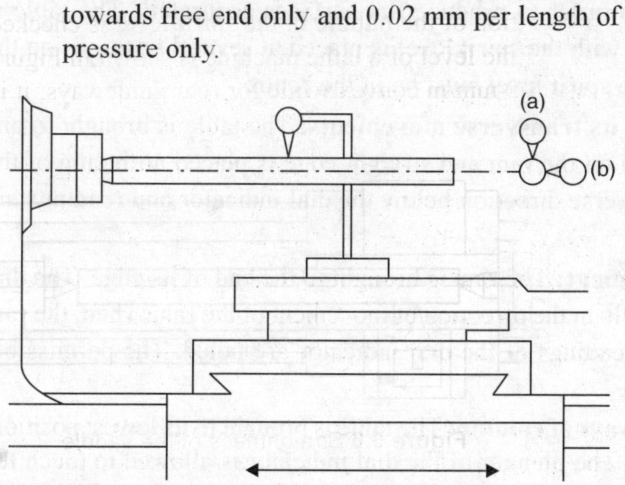

Figure 5.9 Parallelism of main spindle to saddle movement

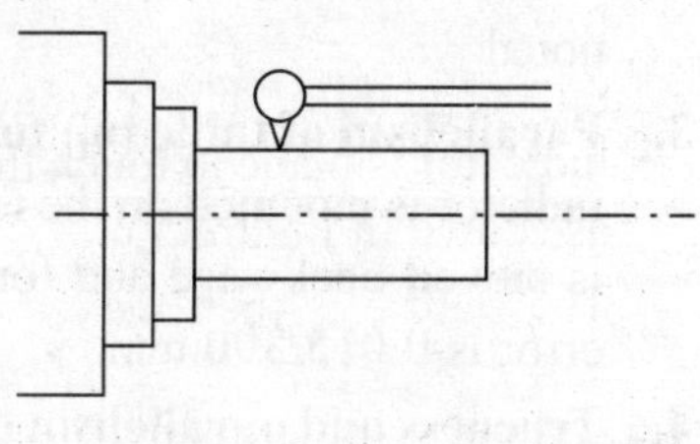

Figure 5.10 True running of locating cylinder of main spindle

6. **True running of locating cylinder of main spindle:** The dial indicator is mounted on the carriage (or any other fixed member) and the plunger of the dial indicator is made to touch the locating cylinder as shown in Figure 5.10. The surface of main spindle is rotated by hand and readings of dial indicator are taken. Permissible error is 0.01 mm.
7. **True running of head stock center:** The plunger of the dial indicator is made to touch perpendicular to the tapered surface of the center as shown in Figure 5.11 and the spindle is rotated. The deviation indicated by the dial indicator gives the trueness of the center.

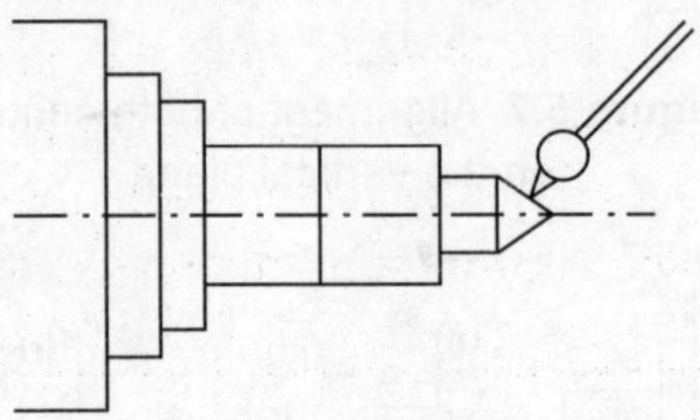

Figure 5.11 True running of head stock center

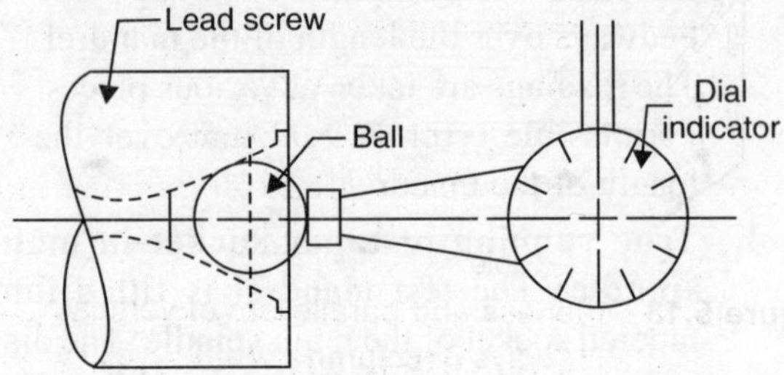

Figure 5.12 Axial slip of lead screw

8. **Axial slip of lead screw:** The thrust face and collars of the lead screw must be exactly square to the screw axis, otherwise a cyclic endwise movement is setup which is of the same nature as the axial slip in the main spindle. In order to test axial slip in the lead screw, a ball is fitted in the end of lead screw, and the plunger of the dial indicator is pressed against the ball as shown in Figure 5.12. The lead screw is rotated and deviation, if any, is noted.

5.6 ALIGNMENT TESTS ON SHAPER

The some of the important alignment tests performed on shaper are described below:

1. **Straightness and flatness of the table:** The straightness and flatness of the table is the most important requirement of the shaper in order to produce flat surfaces accurately. The table is brought in the centre. The gauge block with the spirit level is placed at several positions on the table. At all positions, the bubble of the spirit level must be in the centre.
2. **Parallelism of top surface of table to its transverse movements:** The table is brought to one side end. The dial indicator is mounted on the ram and straight edge is placed at the top of the table. The table is then moved in transverse direction below the dial indicator and readings are noted.
3. **Parallelism of table top to ram movement:** The ram is brought to the end of its edge. The dial indictor is mounted on the top of the table in the direction of movement of the ram. Then, the ram is moved backward and forward and readings of the dial indicator are noted. The permissible error is 0.015/300 mm.
4. **Trueness and parallelism of vertical ways of column:** The table is brought to its lowest position. The dial indicator is fixed on the table. The plunger of the dial indicator is allowed to touch the vertical ways of the column as shown in Figure 5.13. Then, the table is moved up. If the side ways are perfectly parallel and levelled, the dial indicator will not show any deflection.

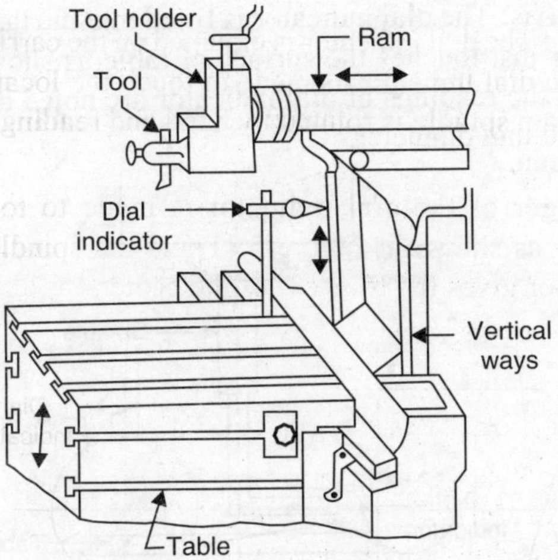

Figure 5.13 Trueness and parallelism of vertical ways of column

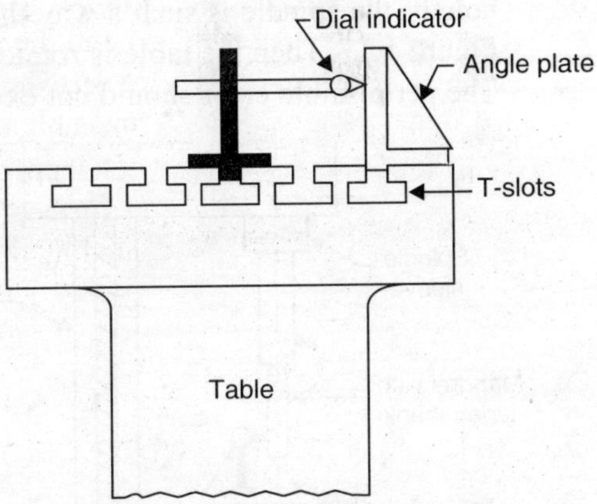

Figure 5.14 Checking accuracy, squareness and parallelism of T-slots

5. **The accuracy, squareness and parallelism of T-slots on the table:** The angle plate is inserted in the slot lengthwise and the dial indicator is mounted in the adjacent parallel slot as shown in the Figure 5.14.

   The plunger of the dial indicator is made to touch the angle plate. The dial indicator is then moved through the slot lengthwise and the deflection, if any, is noted.

5.7 ALIGNMENT TESTS ON PILLAR TYPE DRILLING MACHINE

The some of the important alignment test performed on pillar type drilling machine are as follows:

1. **Flatness of clamping surface of base and table:** The test is performed by placing a straight edge on two gauge blocks placed on the base plate as shown in Figure 5.15. The error is noted down by inserting the feeler gauges. The permissible error should not exceed 0.1/1000 mm and the surface should be concave only.

   The flatness of clamping surface of table is also checked in the same way. Straight edges on two gauge blocks are placed on the table in different positions and the error is noted by inserting feeler gauges. The permissible error is also the same.

2. **True running of spindle taper:** The test mandrel is fitted into the tapered hole of the spindle. The dial indicator is mounted on the table in such a way that its plunger just touches the test mandrel as depicted in Figure 5.16.

   The spindle is rotated slowly and readings of dial indicator are noted. The permissible error is 0.01 mm.

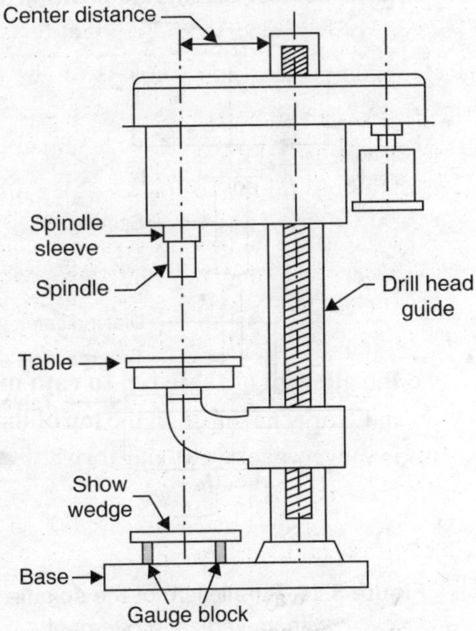

Figure 5.15 Checking flatness of clamping surface of base

3. **Squareness of clamping surface of table to its axis:** The dial indicator is fitted into the tapered hole of the spindle is such a way that its plunger just touches the surface of table as shown in Figure 5.17. Then the table is rotated slowly and the readings of dial indicator are noted down. The permissible error should not exceed 0.05/300 mm diameter.

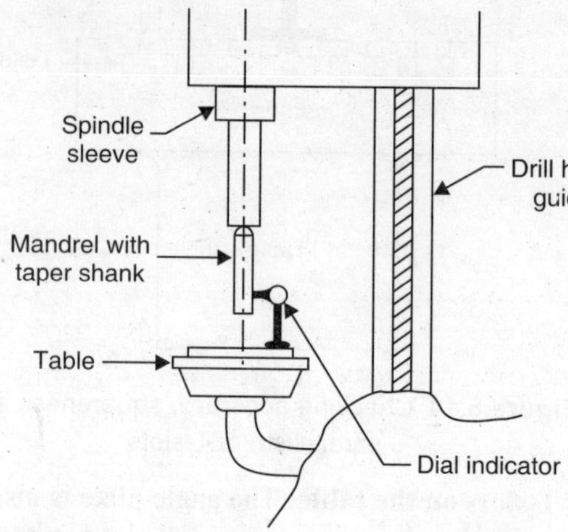

Figure 5.16 True running of spindle taper

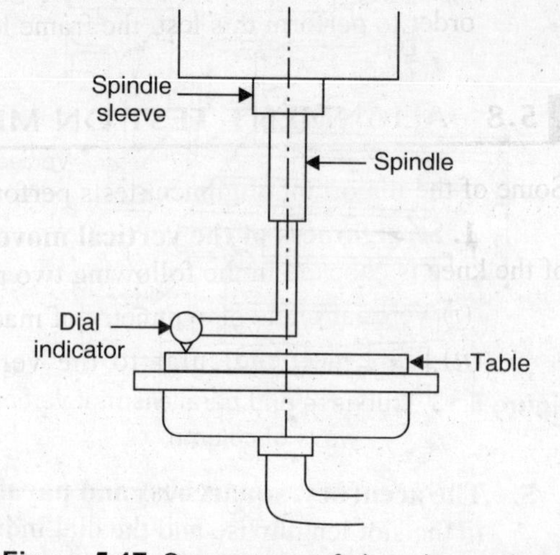

Figure 5.17 Squareness of clamping surface of table to its axis

4. **Parallelism of the spindle axis with its vertical movement:** This test is carried out in two planes (*a*) and (*b*) at right angles to each other as shown in Figure 5.18. The test mandrel is fitted in the tapered hole of the spindle and the dial indicator is mounted on the table in a manner that its plunger just touches the test mandrel. The spindle is adjusted in the middle position of its travel. The readings of the dial indicator are noted when the spindle is moved in upper and lower directions of the middle position with slow vertical feed mechanism.

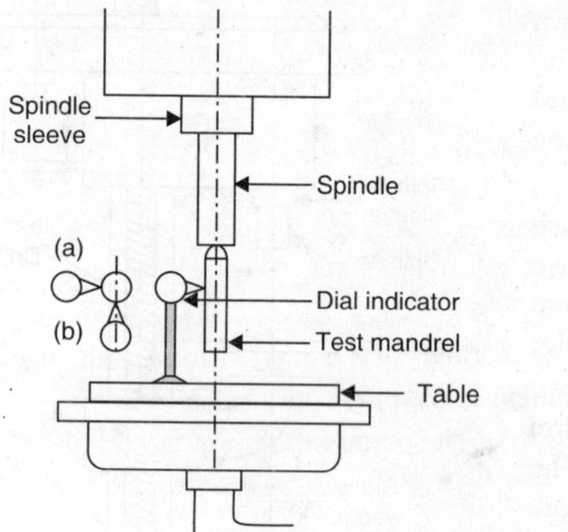

Figure 5.18 Parallelism of the spindle axis with its vertical movement

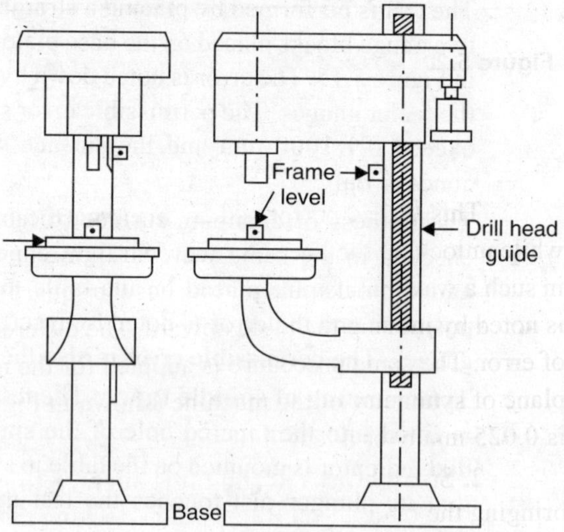

Figure 5.19 Test for perpendicularity of drill head with table

5. **Perpendicularity of drill head guide with table:** The frame level (with graduations from 0.03 to 0.05 mm/m) is placed on guide column and table. The error is recorded by noting the difference between the readings of the two levels. The permissible error is 0.25/1000 mm. In the similar way, perpendicularity of drill head guide with table is also checked as shown in Figure 5.19. In order to perform this test, the frame level is placed on the base instead of table.

5.8 ALIGNMENT TEST ON MILLING MACHINE

Some of the important alignment tests performed on milling machine are as follows:

1. Straightness of the vertical movement of the knee: Straightness of the vertical movement of the knee is checked in the following two planes:

(*i*) Vertical plane of symmetry of machine as depicted in Figure 5.20(*a*).

(*ii*) Plane perpendicular to the vertical plane of symmetry of the machine depicted in Figure 5.20(*b*).

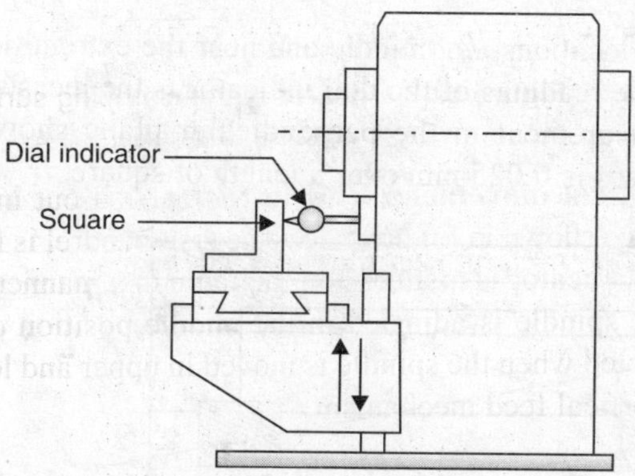

Figure 5.20 (*a*) Straightness of the vertical movement of the knee in the vertical plane of symmetry.

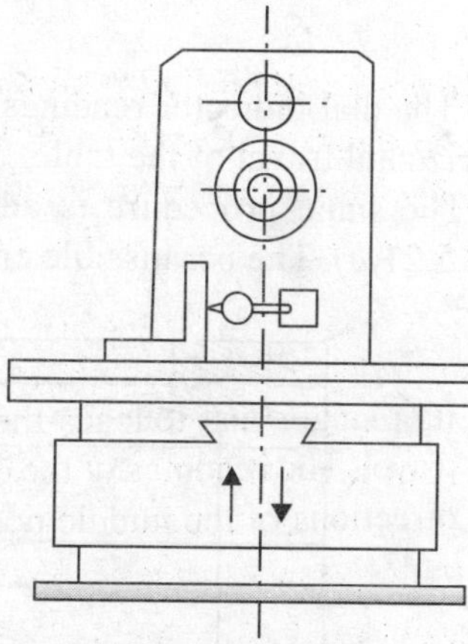

Figure 5.20 (*b*) Straightness of the vertical movement of the knee in the plane perpendicular to the vertical plane of symmetry

This test is performed by bringing the table in central position, locking tabic and cross-slide while unlocking the knee. A square is fixed on the surface of the table and the dial indicator is mounted in such a way that its plunger touches the square as shown in Figure 5.20 (*a*). The dial indicator reading is noted by moving the table upward. The difference in the readings of the dial indicator is the measure of error. The same procedure is adopted for the measurement in the plane perpendicular to the vertical plane of symmetry of the machine (shown in Figure 5.20 (*b*)). The permissible error in both the planes is 0.025 mm over a length of square.

2. Straightness of the table surface to the column ways for knee: This test is performed by bringing the table in central position, locking table and cross-slide while unlocking the knee. A square is fixed on the surface of the table and the dial indicator is mounted in such a way that its plunger touches the square as shown in Figure 5.21(*a*).

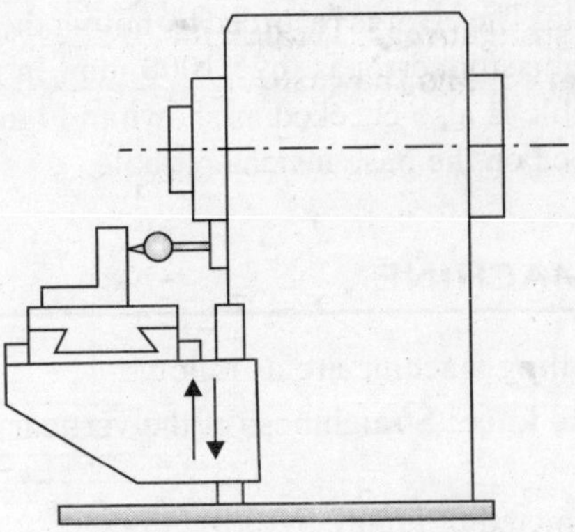

Figure 5.21 (*a*) Straightness of the table surface to column ways for knee in the vertical plane of symmetry

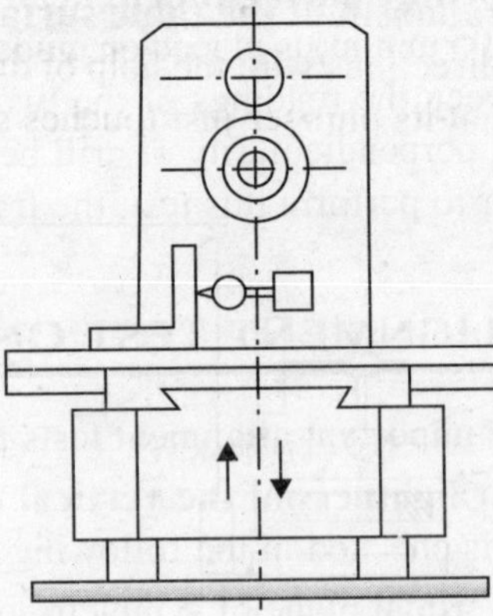

Figure 5.21 (b) Straightncss of the table surface to column ways for knee in the plane perpendicular to the vertical plane

The dial indicator readings are noted at three locations *i.e.*, middle and near the extremities of the horizontal travel of the table. The difference in the readings of the dial indicator is the measure of error. The same procedure is adopted for the measurement in the perpendicular plane shown in Figure 5.21(*b*). The permissible error in both the planes is 0.025 mm over a length of square.

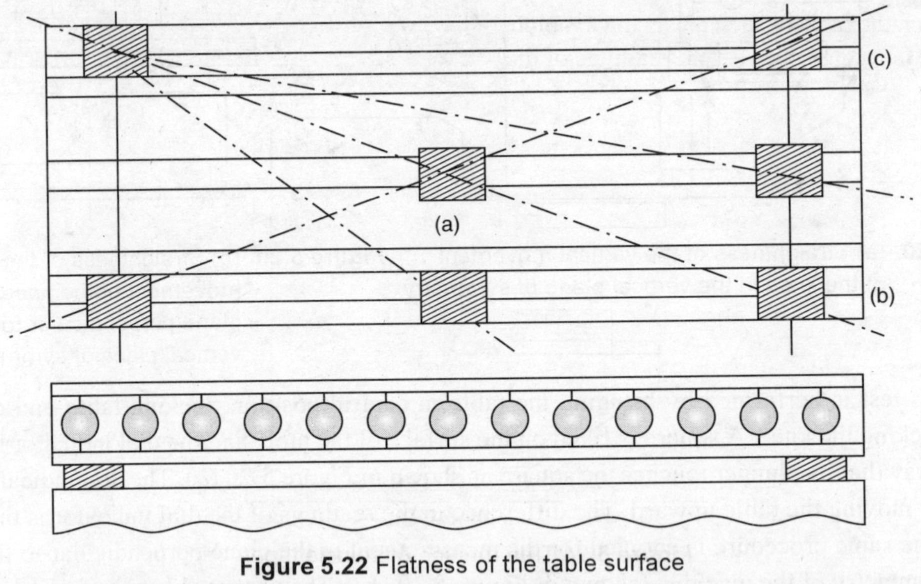

Figure 5.22 Flatness of the table surface

3. Flatness of the table surface: The flatness of the table surface is important requirement of the milling machine. It is checked with the help of spirit level. This test is performed bringing table and cross-slide in its central position. The table is unlocked while knee and cross-slide are locked. The spirit level is placed at several positions on the table as shown in Figure 5.22. At all positions, the bubble of the spirit level must be in its central position. Maximum permissible error is 0.05 mm.

4. Parallelism of the table surface to its movement: This test is performed in transverse and longitudinal directions with the help of dial indicator and straight edge. The dial indicator is mounted in such a way that its plunger just touches surface of straight edge as shown in Figure 5.23.

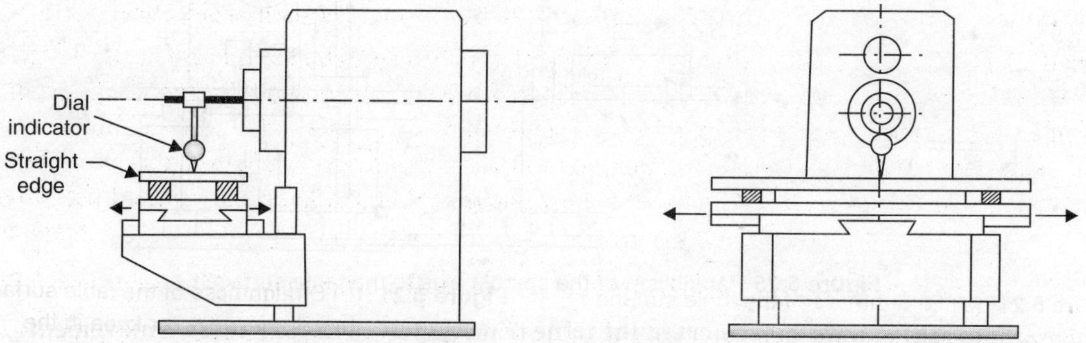

Figure 5.23 Parallelism of the table surface to its movement in transverse and longitudinal direction.

Parallelism of the table surface in transverse direction is checked by locking table and spindle head slide while parallelism in longitudinal direction is checked by locking cross slide and spindle head slide. Maximum permissible error in both directions is 0.02 mm over the length of square.

5. True running of spindle taper: This test is performed by arranging the table in its main position longitudinally and the test mandrel is fitted into the tapered hole of the spindle. The dial indicator is mounted on the table in such a way that its plunger just touches the test mandrel as shown in Figure 5.24. The spindle is rotated slowly and readings of dial indicator are noted. The permissible error is 0.01 mm.

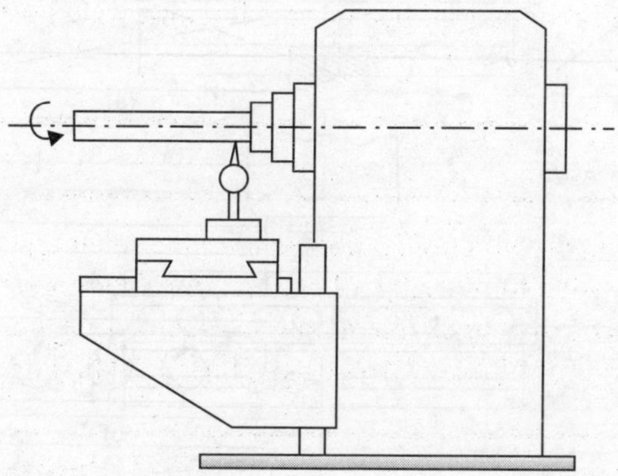

Figure 5.24 True running of spindle taper

6. Parallelism of the spindle axis to the table surface: This test is performed in two planes by locking table, spindle head slide, cross slide and knee. The test mandrel is fitted in the tapered hole of the spindle and dial indicator is mounted on the machine table. The plunger of dial indicator just touches the surface of test mandrel as shown in Figure 5.25.

The readings at the maximum vertical travel of the table surface are noted at two positions. The permissible error in both planes is 0.025 mm over the length of test mandrel.

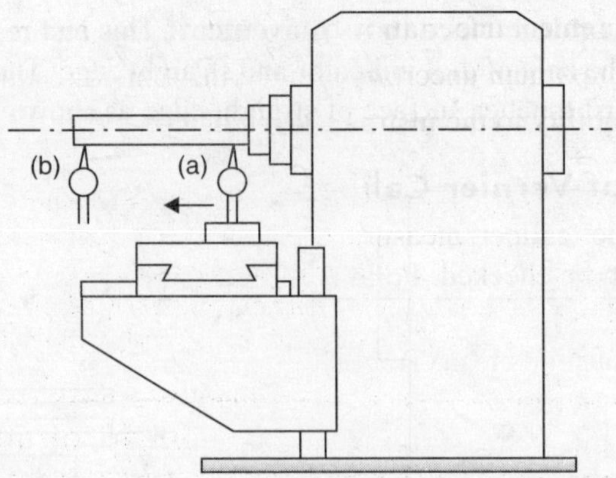

Figure 5.25 Parallelism of the spindle axis to the table surface

7. Squareness of the movement of the table transversely to its longitudinal movement: This test is performed by arranging the straight edge parallel to the longitudinal movement of the table. The square is then placed against the straight edge. The dial indicator is fitted as shown in Figure 5.26. The table is locked in its central. position and dial indicator readings are noted. The permissible error is 0.02 mm over the length of square.

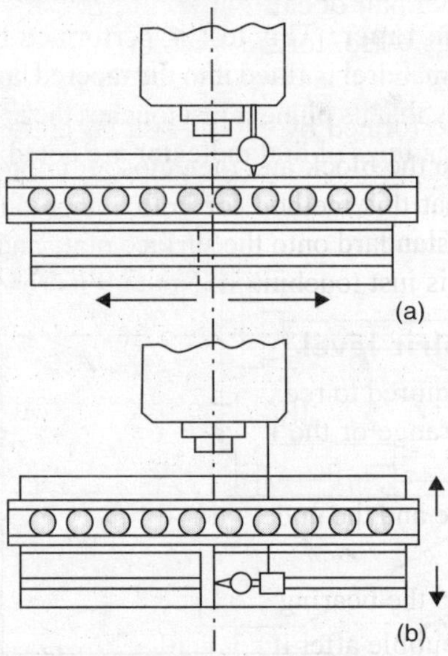

Figure 5.26 Squareness of the movement of the table transversely to its longitudinal movement.

5.9 RECALIBRATION OF VARIOUS MEASURING INSTRUMENTS

In order to check and restore the accuracy of measuring instruments, recalibration is carried out. The accuracy of the instruments drifts with the passage of time and environmental conditions. To maintain specified measurement accuracy, instruments must be recalibrated at regular intervals as specified by the manufacturer. The significance of recalibration is as given below:

(*i*) It traces the measurement uncertainty of the instruments and restores back its accuracy.

(*ii*) It determines the maximum uncertainty in the instrument reading.

(*iii*) It provides acceptability to the instrument reading.

5.9.1 Recalibration of Vernier Caliper

Before recalibrating the vernier caliper, measuring surfaces must be thoroughly checked for scratches. Scale of the caliper must also be checked. Following steps must be followed for recalibrating vernier caliper:

1. The caliper jaws must be clean. Track of grease, dirt or other foreign particle must be clean with clean cloth.
2. After closing the caliper jaws and readings must be noted down. The dial should read zero. If it doesn't read zero, the dial must be adjusted to zero manually.
3. Check the OD jaws first by inserting a 0.5 inch gauge block in between the jaws. Close the caliper, so that both the jaws are touching the block. Note this reading. Repeat this step with a 1 inch gauge block, and then with a 4 inch gauge block. Note down these readings. Then, repeat the same procedure using the appropriate size micrometer standard (150, 200 or 300) and note these readings.
4. Check the ID jaws by setting the calipers to 0.5 inch and locking them into place. Measure the jaw width with another pair of calibrated calipers. Note this reading. Repeat this step for 1 inch and 4 inch settings. Also, for the appropriate size micrometer standard (150, 200 or 300). Note these readings.
5. Height check must be performed by setting a 0.5 inch gauge block onto the surface plate. Extend the caliper above the block and then close it until it is just touching the block. Note down the reading. Repeat this method for both 1 inch and 4 inch gauge blocks. Place the correct size micrometer standard onto the surface plate and extend the caliper above it. Then close the caliper until it is just touching it. Note all its readings.

5.9.2 Recalibration of Spirit level

An adjustable bearing surface is required to recalibrate the spirit level. It may not be in perfect level to start with, but must be within the range of the level. Following steps must be taken for recalibrating spirit level:

1. Clean the bearing surface and the underside of the spirit level to remove any dirt, dust or any other foreign particle.
2. Place the spirit level onto the bearing surface.
3. Note the position of the bubble after it has settled down and then turn the spirit level by 180° and put onto the surface in the same position as before. Again note the position of the bubble after it has settled.
4. If the spirit level and the bearing surface are set in level, then the bubble will be in its central position in both cases.
5. If the spirit level is set in level but the bearing surface is out of level, then the bubble will move in the direction off-centre when the spirit level is turned by 180°. The surface can then be readjusted by this amount.
6. On the other hand, if the bearing surface is set in level but the spirit level is out of level, then the bubble will move in different direction off-centre when the level is turned by 180°. The spirit level must be then readjusted to read level.

5.9.3 Recalibration of Dial Indicator

Dial indicator must be calibrated before using to ensure readings are correct and its measurements are accurate. Following steps must be taken while recalibrating dial indicator:

1. Surface plate, gauge blocks and the plunger of dial indicator must be cleaned to remove any dust or dirt.
2. Dial indicator should be locked in place on its stand, resting high enough to allow the plunger to travel its full length and should be able to just touch the surface plate.
3. Touch the plunger on the surface plate and rotate the face of dial indicator until the "0" marking aligns with the dial.
4. Raise the plunger and slip a 0.020 inch gauge block underneath it. Lower the plunger onto the surface of the gauge block. Check that the indicator reads 0.020 inch. Repeat this thrice and check that the dial indicator gives the same readings.
5. Repeat the step with 0.025 inch, 0.05 inch, 0.1 inch, 0.25 inch, 0.4 inch and 1 inch gauge blocks. Check each block twice to ensure the dial indicator's repeatability.

SUMMARY

- Correct installation is the first essential step before testing a machine tool for various acceptance tests.
- Alignment and performance tests are collectively known as acceptance tests.
- Alignment tests deal with manufacturing accuracies while performance test deal with the working accuracies.
- The manufacturer usually provides a manual along with the machine tool, containing information regarding the foundation plan, guidance for installation and the test charts.
- The test charts provide a sort of guarantee that the machine tool has been tested in such conditions under which it is required to work.
- Practical tests for checking accuracy and surface finish of the product being machined may also be carried out.
- Dial indicator is used for alignment tests to check parallelism and concentricity of holes or rods and flatness of surfaces 10 an accuracy of 0.01 mm.
- To check the true running of the spindle, test mandrels are used.
- Spirit level is used for leveling the machine.
- Straight edge is placed on machined surface to check them for flatness or straightness.
- The straightness and flatness of the table is the most important requirement of the shaper in order to produce flat surfaces accurately.
- The accuracy of the measuring instruments drifts over time, with factors like time in service and environmental conditions. To maintain specified measurement accuracy, instruments must be re-calibrated at regular intervals as defined by the manufacturer.
- Re-calibration traces the measurement uncertainty of measuring instruments and restores back its accuracy.

ASSIGNMENT

1. **Students should be asked to perform various alignment tests on lathe, shaper, drilling machine and milling machine and to prepare test charts.**

EXERCISES

1. What is meant by alignment tests on machine tools? Why are they necessary? Explain.
2. List the various geometrical checks made on machine tools.
3. Distinguish between geometrical and practical tests on a machine tool.
4. Name the various instruments required for performing the alignment test on machine tools.
5. Describe the following instruments with neat sketches:
   (*i*) Dial gauge (*ii*) Test mandrel
   (*iii*) Spirit level (*iv*) Straight edges and squares.
6. Name the various alignment tests to be performed on a lathe. Describe any two of them in detail.
7. Name the various alignment tests to be performed on a shaper. Describe any two of them in detail.
8. Describe the following alignment tests as referred to shaper:
   (*i*) Straightness and flatness of the table.
   (*ii*) Parallelism of top surface of table to its transverse movements.
   (*iii*) Parallelism of table top to ram movement.
   (*iv*) Trueness and parallelism of vertical ways of column.
9. Explain with neat sketches various alignment tests that are performed on pillar type drilling machine.
10. Describe the following alignment tests as referred to pillar type drilling machine:
    (*i*) Flatness of clamping surface of table.
    (*ii*) Perpendicularity of drill guide to the table.
    (*iii*) Parallelism of the spindle axis with its vertical movement.
11. With the neat sketches, explain various alignment tests that are performed on vertical knee type milling machine.
12. Describe the following alignment tests performed on vertical knee type milling machine
    (*i*) Straightness of the vertical movement of the knee
    (*ii*) Straightness of the table surface to the column ways for knee.
    (*iii*) Flatness of the table surface.
    (*iv*) True running of spindle taper.
13. What is the significance of re-calibration of measuring instruments?
14. Describe the steps followed for recalibrating vernier caliper.
15. How spirit level can be recalibrated?
16. Describe the steps followed for recalibration of dial indicator.

❑❑❑

Lubrication

6

CHAPTER

6.1 LUBRICATION

The science which deals with reduction of friction by application of lubricant between the rubbing surfaces of bodies having relative motion is known as lubrication. When the lubricant is applied into the clearance which prevents meeting and engaging of surfaces, it decreases the frictional forces and minimizes the wear and tear. Thus, a lubricant reduces the waste of energy, surface wear, damage, deformation and operational costs.

Lubrication is essential for correct operation of mechanical systems like pistons, pumps, cams, bearings, turbines, cutting tools, etc., as without lubrication the pressure between the surfaces in close proximity generates enough heat which rapidly causes surface damage, in a coarsened condition which may cause seizure literally by welding the surfaces together. In some applications, like the piston engines, the film between the piston and the cylinder wall seals the combustion chamber which may prevent escaping combustion gases.

6.2 FUNCTIONS OF LUBRICATION

Lubrication performs the following major functions:

(*i*) Minimization of the frictional power loss.

(*ii*) Minimization of the wear of the rubbing parts.

(*iii*) Minimization of rusting and corrosion.

(*iv*) It works as coolant.

(*v*) It also keeps machine parts clean and free from dirt.

6.3 CLASSIFICATION OF LUBRICANTS

We can easily classify lubricants in the following manner:

(*i*) Solid lubricants: Graphite, molybdenum disulfide, mica, chalk, talc, boron nitride, polytetrafluorethylene (PTFE), calcium fluoride, tungsten disulfide, etc.

(*ii*) Semi-solids lubricants: Grease, vaseline, etc.

(*iii*) Liquid lubricants: Mineral oils (Petroleum fractions), animal oils (Whale oil, Lard oil and Tallow oil), vegetable oils (Olive oil, Palm oil), synthetic oils (Silicones, Fluoluboes).

(*iv*) Emulsions like oil in water type (cutting emulsions), water in oil type (cooling liquids).

6.4 DESIRED CHARACTERISTICS OF LUBRICANTS

A good lubricant should have the following characteristics:

(*i*) High viscosity

(*ii*) High surface tension.

(*iii*) High flash and fire points.

(*iv*) Good oiliness and adhesiveness.

(*v*) Low pour point.

(*vi*) Low volatility.

6.5 PROPERTIES OF LUBRICANTS

The important properties of a lubricant are discussed below:

(*i*) **Viscosity:** The measure of degree of fluidity of the oil and internal resistance offered against shear of the oil film is called viscosity. Heavier loads are supported by high viscosity oil which has more internal friction. As temperature of oil increases, it viscosity decreases. Thus, the proper operation of equipment depends on the proper viscosity at high operating temperature.

(*ii*) **Surface tension:** It is the property of a liquid to cause bubbles and droplets due to adhesive forces among the liquid molecules at the surface of the liquid.

(*iii*) **Flash and fire points:** Flash point is the minimum temperature at which application of a flame causes specimen vapors to ignite momentarily. Flash point is used to assess the overall hazard of a material and is used in handling, transportation and storage of oil. Fire point is the minimum temperature at which application of a flame causes specimen vapors to burn continuously for at least 5 seconds.

(*iv*) **Oiliness and Adhesiveness:** Oiliness is the property of the lubricant by virtue of which oil can stick on to the surface of the machine parts operating under high pressure. It is a joint property of oil and metal surface in contact. There is no absolute measure for oiliness. The value of oiliness is only comparative by referring the oiliness curve. For example, lard oil has better oiliness than mineral oils. Adhesiveness is the ability of the lubricant to stick to the metal surface to obstruct metal to metal contact.

(*v*) **Pour point:** The lowest temperature at which lubricating oil ceases to flow when cooled is known as pour point. It indicates the minimum temperature at which oil may be used. In cold temperatures, the lubricant having low pour point is more useful.

(*vi*) **Volatility:** It is the ability of the liquid to vaporize at a high temperature. Evaporation loss is particularly important to motor and cylinder lubrication, due to the high operating temperature and the evaporative loss to increases at high temperatures. Significant evaporative loss of oil leads to excessive oil consumption and drastic changes in oil properties.

6.6 SELECTION OF LUBRICANTS

(*i*) **Manufacturer recommendations:** The film thickness and wear are of utmost importance. If we want to test lubricants, they are subjected to various types of physical stress. Experience has probably played a larger role than any other single criterien. Testing and experience has taught the machine manufacturers which lubricants are preferred for the better care of their products.

Professional societies have classified and established specific lubricants preferable in a given mechanical situation. For example, AGMA has given standard specifications for gear systems. Along with a specific brand, they provide sometimes a list of alternative lubricants, for their machines.

(*ii*) **Recommendations of lubricant manufacturers:** The manufacturers usually recommend lubricants for their products in terms of specifications or required qualities rather than particular brand names, the user has to identify the useful brands. In case a user is unsure lubricant manufacturer may be consulted on products that comply with the required specifications. On the basis of specific product, operating characteristics, unusual operating requirements, and lubricant specification, product engineers may easily identify lubricants conducive to the manufacturer's specifications.

(*iii*) **User selection:** Applicable criteria should be the basis of lubricant selection made by the user. Selection should be in the class recommended by the machine manufacturer and be in the same base stock category (paraffinic, naphthenic, or synthetic). Moreover, the physical and chemical properties should be comparable with the manufacturer's specification.

6.7 SEMI-SOLID LUBRICANT-GREASE

Greases are applied in those areas where a continuous supply of oil cannot be maintained, such as in open bearings or gears. Factors to be considered while selecting suitable grease are operating temperatures, water resistance, oxidation stability, etc. Another important factor is the characteristics of the grease which include viscosity and consistency.

Lubricating grease is a mixture of base oil, performance additives and a thickener forming a matrix retaining the oil in a semisolid state. Most grease thickeners are soaps, i.e. lithium, calcium, or aluminum soap. Complex soap greases have superior temperature resistance. They are commonly usable up to 180°C at which the mineral oil vaporizes. A smaller number of greases, restricted to very special applications, are manufactured with non-soap thickeners such as organoclays, polyurea or silica compounds.

Grease is the most commonly used lubricant for roller bearings and low velocity applications, because grease type lubricants are relatively easy to handle and need the simplest sealing devices.

6.8 MODE OF LUBRICATION

Thick and thin film lubrications are the basic modes of lubrication.

6.8.1 Thick Film Lubrication

Thick film lubrication is a condition of lubrication, in which two surfaces of the bearing in relative motion are completely separated by lubricant. The resistance to relative motion erupts from the viscous resistance of lubricating oil. Thus, the viscosity of lubricating oil influences the performance of bearing. The thick film lubrication may be again divided into two categories:

(*i*) Hydrodynamic lubrication

(*ii*) Hydrostatic lubrication

Hydrodynamic Lubrication

It is a system of lubrication in which the load supporting film is created by the shape and relative motion of the sliding surfaces. The principle of hydrodynamic lubrication in journal bearing is shown in Figure 6.1. The shaft is rest (*a*) and it sinks to the bottom of the clearance space under the action of load W. The journal and bearing surfaces touch during 'rest'. When the journal starts rotating, it climbs the surface of bearing (*b*) and its speed further increases; it forces the lubricating oil into the wedge-shaped region (*c*). As more and more oil is forced in the wedge-shaped clearance space, pressure is created within the system.

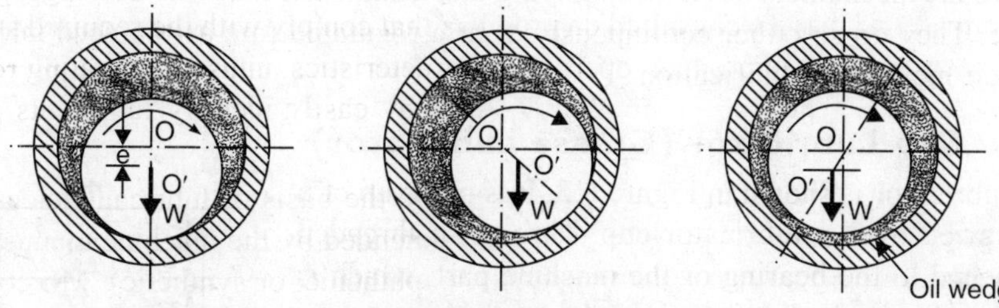

Figure 6.1 Hydrodynamic Lubrication

Hydrostatic Lubrication

It is a system of lubrication in which, the load supporting fluid, separating the two surfaces is generated by an external source, as the pump, supplying sufficient fluid under pressure. The hydrostatic lubrication in journal bearing is depicted in Figure 6.2.

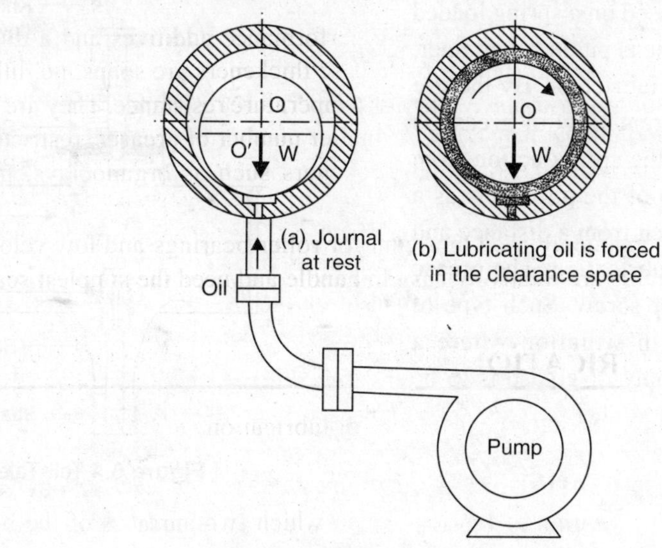

Figure 6.2 Hydrostatic Lubrication

In the beginning, the shaft is at rest on the bearing surface. (*a*) When the pump starts, high-pressure lubricating oil is forced in the clearance space, which forces the surfaces of bearing and journal to separate (*b*). The vertical turbo generators, ball mills uses this types of lubricating system, etc.

6.8.2 Thin Film Lubrication

Thin film lubrication, which is also called boundary lubrication, is a condition of lubrication with a relatively thin lubricant film and there is partial metal-to-metal contact. This mode of lubrication can be noticed in door hinges and machine tool slides. The conditions of boundary lubrication may be due to excessive load, insufficient surface area or oil supply, low speed and misalignment.

6.9 LUBRICATORS

The following are the mechanical devices, supplying lubricant to the machine bearings, gears and other engine parts, etc. They do so either continuously or at a regulated rate as the need may be. They are named as per their methods of lubrication.

6.9.1 Screw Cap Lubricator (Grease Lubricator)

The screw cap lubricator is shown in Figure 6.3. It is made up of a knurled cap screwed over a lubricator cup. Through a threaded nipple, it is attached to the bearing or the machine part, which is tightened over the place to be lubricated. Under manual operation, its process includes turning the cap and forcing the grease out on the surface to be lubricated through the hole. This type of lubricator is most useful in food or chemical processing equipments, with heavily loaded parts.

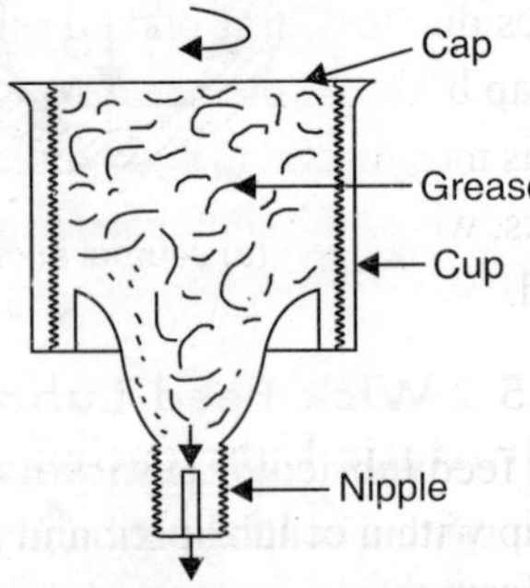

Figure 6.3 Screw Cap Lubricator

6.9.2 Tell-tale Lubricator

Figure 6.4 shows Tell-tale lubricator. It feeds the grease by spring pressure based on a spring loaded piston. In this type, the grease is put in a container, screwed on the body of the lubricator. By the use of slotted grub screw, different flow of grease is managed. The piston stops the spring action from emptying the cap. The knob of the piton acts as a "Tell-tale", which can be seen from a distance and indicating the progress of lubrication. The grease is supplied by the container screw. Such type of lubricator is most suited in situation where a regulated or continuous supply of grease is to be maintained.

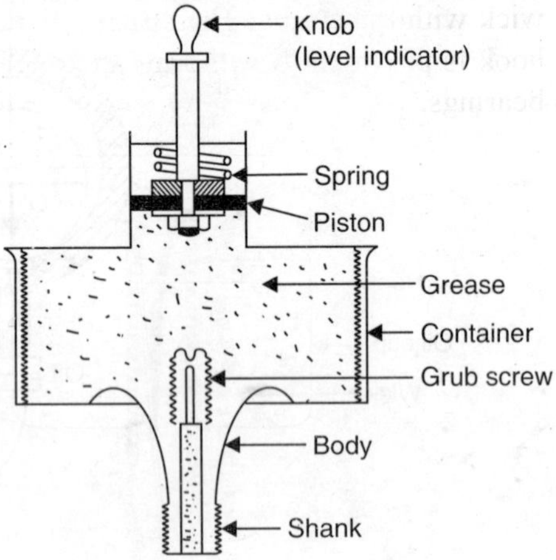

Figure 6.4 Tell-Tale Lubricator

6.9.3 Drop Feed Oil Lubricator

Drop feed oil lubricator is shown in Figure 6.5. It operates in a glass. Oil to the cup is filled by an oil hole at its top. Through the opening the oil flows and falls drop by drop due to gravity through the needle valve. The drops of oils can be seen through the glass window. The lubricator is mounted over the part to be lubricated using tits screw end. This method is best suited to antifriction bearings and high grade machines.

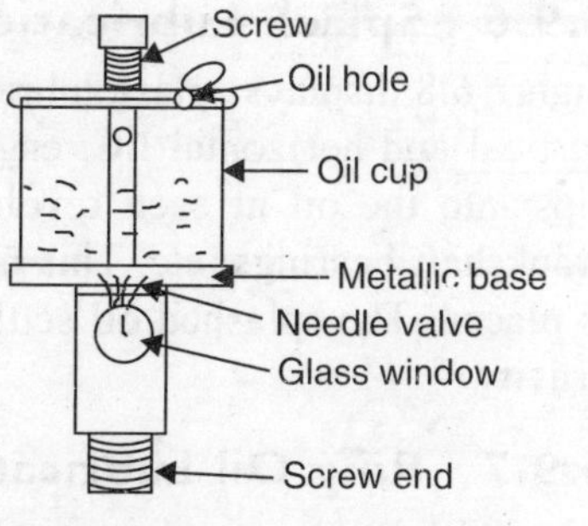

Figure 6.5 Drop Feed Lubricator

6.9.4 Glass Bottle Lubricator

Figure 6.6 shows the Glass bottle lubricator. In this type of lubricator, a glass bottle is filled with oil with a tapered wooden cork which carries a loose fitting steel needle. The whole is reversed and the lower portion of the cork is fitted into the bearing cap so as the needle to rest over the journal. With the rotation of journal, the needle shakes due to vibration and oil from the bottle through the gap between the needle and the cork to the bearing. This is most useful in plain journal bearings and plumber blocks, where the shaft carries moderate loads at average speed.

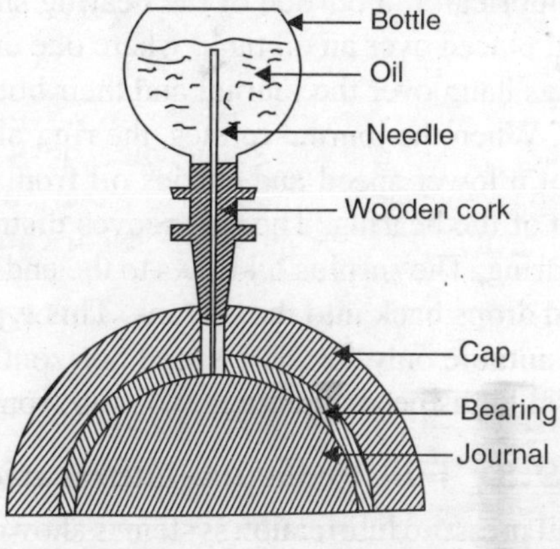

Figure 6.6 Glass Bottle Lubricator

6.9.5 Wick Feed Lubricator

Wick feed lubricator is shown in Figure 6.7. It is works on the siphon principle. It is made up of a glass oil cup with a central pipe and a threaded shank, screwed in the oil hole. In this type of lubricator, wick runs from the oil cup to about 50 mm below the lowest level of the oil in the cup. By the capillary action, the wick gets soaked in the oil. Due to gravity and vibration, the oil drip off from the lower end of the wick within the tube. The strands of the yarn regulate the rate of flow. When the machine is idle the hook is provided to withdraw the wick. This is of immense use on guides, slides, spindles and plain bearings.

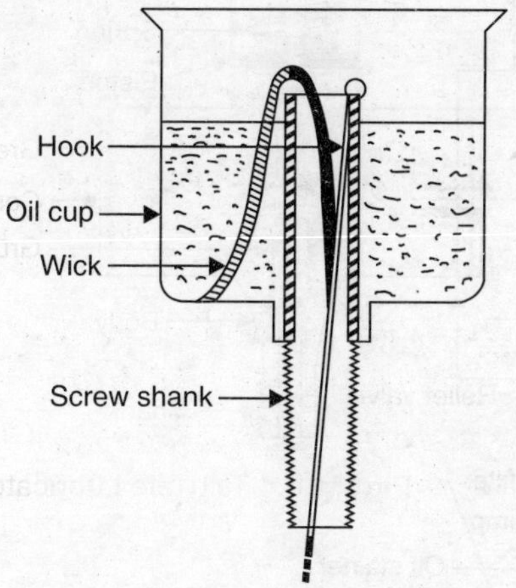

Figure 6.7 Wick Feed Lubricator

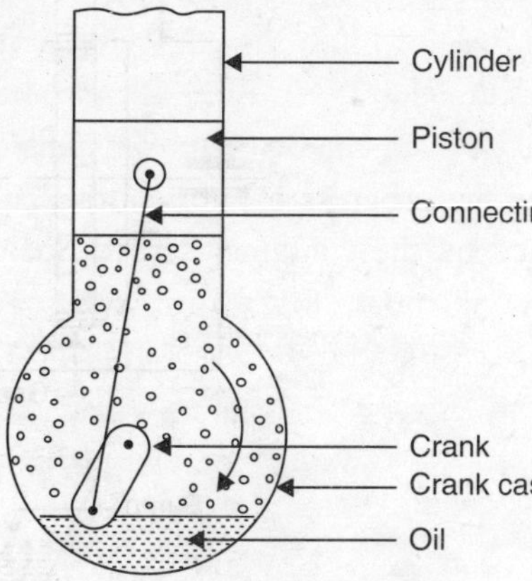

Figure 6.8 Splash Lubricator

6.9.6 Splash Lubrication

Figure 6.8 displays splash lubricator, which is used to lubricate the cylinder and piston in high-speed vertical and horizontal I.C. engines. The closed crankcase contains oil at a level that the crank pin dips into the oil at each revolution and splashes the surfaces of the cylinder, piston, piston ring, crankshaft bearings, etc. This method is useful for gears, chains and bearings over which the trough is placed. The splashed oil settles in the trough which flows to the parts and lubricates them due to gravity.

6.9.7 Ring Oil Lubricator

Ring oil lubricator is shown in Figure 6.9. In this type of lubricator, a portion of the bearing shell is cut away and placed over an oil tank, where one or more metallic rings hang over the journal and their bottom dips in the oil. When the journal rotates, the ring also rotates with it at a lower speed and carries oil from the bath to the top of the bearing. The oil grooves distribute oil to the bearing. The surplus oil flows to the ends of the bearing and drops back into the oil tank. This type of lubricator is suitable only for slow speed horizontal line shafting as at high speeds the rings will slip from bearing on account of the centrifugal force.

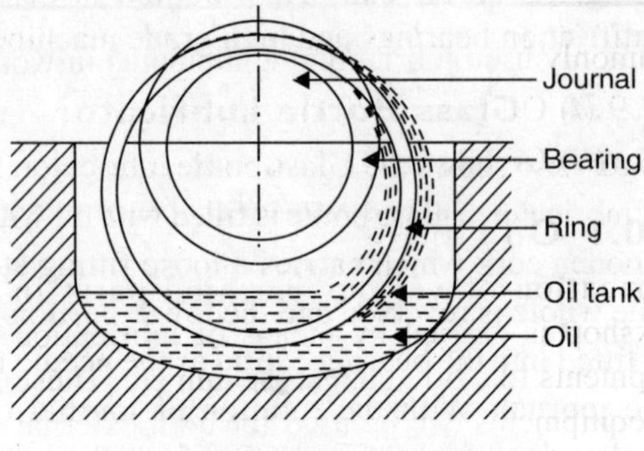

Figure 6.9 Ring Oil Lubricator

6.9.8 Full Pressure Lubrication

Full pressure lubrication system is shown in Figure 6.10. In this method, the pump is used for circulation of oil to different lubricating zones. The oil is taken from the sump through the strainer. The pump consists of a pair of gears. One gear gets its motion from the crankshaft or camshaft whereas the other gear is inoperative.

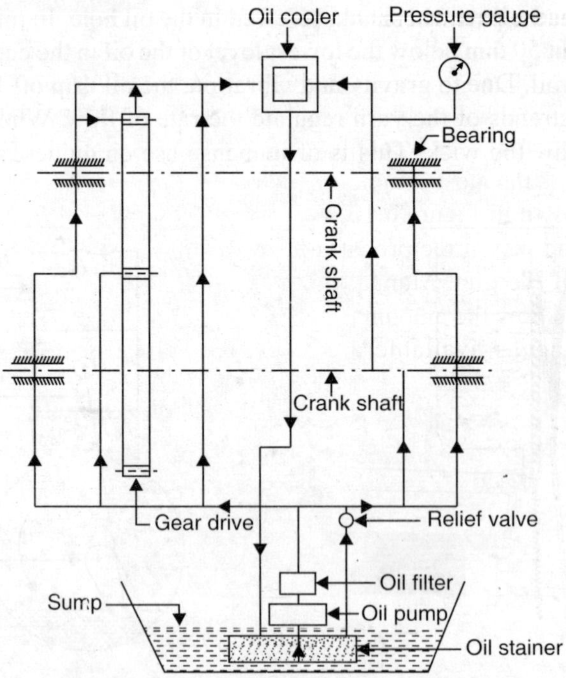

Figure 6.10 Full Pressure Lubrication

The oil drawn by the pump is further forced through the oil filter. When the pressure of the oil exceeds normal pressure, the oil returns to the sump through the relief valve. The oil reaches to the oil cooler through a set of tubes and cold water is passed around them. From the oil cooler the oil goes to the lubricated places from where the excess oil descends into the sump to be recirculated. These are most suited for the heavily loaded bearings.

6.10 COMMONLY USED LUBRICATION EQUIPMENTS

Commonly used lubrication equipments in workshop and garage are:

(*i*) Oil gun

(*ii*) Grease gun

6.10.1 Oil Gun

Lubrication of various machine parts in workshop is carried by means of lubrication equipments like oil gun, grease gun etc. These two equipments can be used for both external as well as internal lubrication. By means of these equipments, the lubricant can be easily introduced to those interior parts of machines, where the operator cannot reach easily. Oil gun is shown in Figure 6.11. It consists of oil pipe, lever, oil reservoir and handle. The lubricating oil is filled in oil reservoir and it is pumped out by the help of lever. Oil comes out through oil pipe. This oil pipe is inserted at the part where lubrication is required.

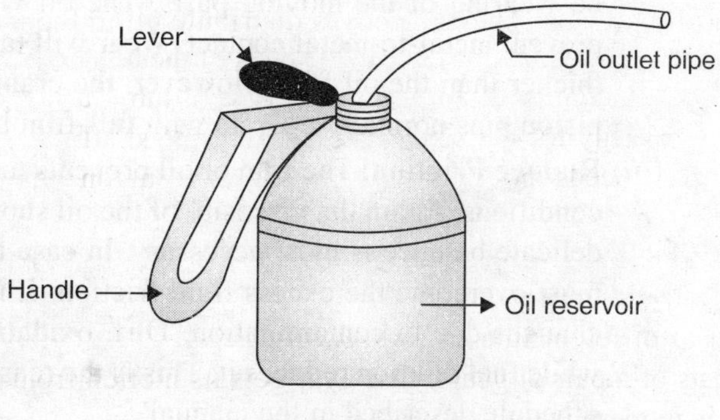

Figure 6.11 Oil Gun

6.10.2 Grease Gun

Greases are applied in those areas where a continuous supply of oil cannot be maintained, such as in open bearings or gears. Grease is the most commonly used lubricant for roller bearings and low velocity applications. Grease gun is shown in Figure 6.12. It consists of cylinder, piston, spring, lever etc. The grease filled in the cylinder and piston reciprocates inside the cylinder. Piston develops low pressure which forces the grease out of the gun. Manual grease guns have disadvantages of over and under lubrication. Technician should know the amount of the grease coming out per stroke of the grease gun. The different types of grease guns available are lever type, pistol-grip type, hand grip type and air-powered etc.

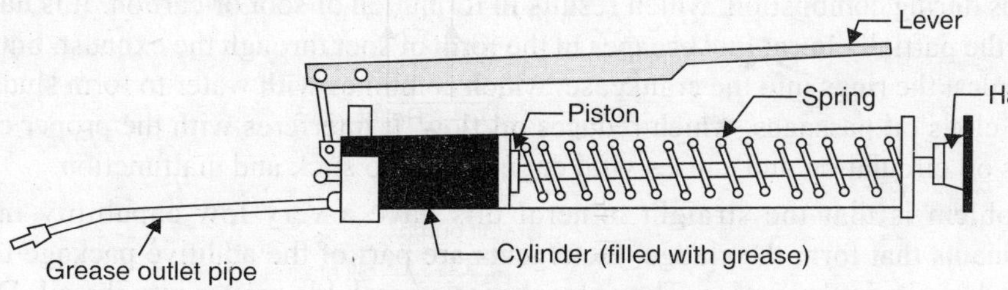

Figure 6.12 Grease Gun

6.11 FUNCTIONS OF ENGINE OIL

Lubricants are mainly used by the automobile industry. Efforts are perpetually made to improve engine designs so as to reduce weight, increase fuel economy, increase power output and at the same time meet environmental emission guidelines. Rigorous research is continuing to formulate lubricants so as to fulfill the needs of the innovated and reshaped engines. The engine oil performs the following major functions:

(*i*) **Permit Easy Starting:** For the first starting the engine to the engine oil should be sufficient thin as to allow a sufficient cranking speed. The oil must flow then be able to flow immediately to lubricate vital engine components. Most of the engine wear takes place at start-up before the oil can reach all the parts of engine. When the engine is heated, the oil must not become too thin so as to be unable to provide engine lubrication adequately.

(*ii*) **Lubricate and Prevent Wear:** With the start of the engine, the oil is being circulated to its parts by the oil pump. The oil must prevent the metal-to-metal contact otherwise there will be wearing of the moving parts. The oil viscosity must remain sufficiently high so as to prevent metal-to-metal contact. Wear will take place, if the surface is scratched by particles thicker than the oil film. However, the crankshaft bearings, connecting rods, camshaft and piston pins normally operate with full-film lubrication.

(*iii*) **Reduce Friction:** The film of oil prevents metal-to-metal contact under full-film lubrication conditions. Again the viscosity of the oil should be high enough to maintain the film. Thus, a delicate balance is most necessary. In case the viscosity is higher than required, the engine must overcome the excess fluid friction. It is important to note that the viscosity of the oil changes due to contamination. Dirt, oxidation and sludge increase the viscosity of the oil while fuel dilution reduces it. This is the reason why the oil must be changed according to the schedule described in the manual.

(*iv*) **Protect Against Rust and Corrosion:** The fuel burns under perfect conditions to form carbon dioxide and water. This water normally escapes as a vapor out of the exhaust, but some amount condenses on the cylinder walls. Water also pass by the piston rings and is trapped in the crankcase. In cold weather, this problem gets worse. Therefore, alongwith water, other corrosive combustion gases also get passes the rings, and they are dissolved in the crankcase oil. The matter becomes worse with the acids formed by the normal oxidation of oil. Thus, the potential for rust and corrosive engine deposits become a hard reality. Corrosion inhibitors as part of the additive package protect non-ferrous metals by coating them by forming a barrier between the parts and the acids. This includes the rust inhibitors which are added to the oil to protect iron/steel surfaces from oxygen attack by forming a protective screen.

(*v*) **Keep Engine Parts Clean:** In petrol or diesel engine due to multiple reasons, the fuel is not completely burned. Some of the partially burnt petrol or diesel undergoes complex chemical reactions during combustion, which results in formation of soot or carbon. It is happens that most of the partially burnt fuel escapes in the form of soot through the exhaust, but some part escapes past the rings into the crankcase, which combines with water to form sludge. Sludge buildup clogs oil passages, which reduces oil flow. It interferes with the proper clearances, restricts oil circulation and causes vital engine parts to stick and malfunction.

The problem is that the straight mineral oils have a very low capability in reducing contaminants that form the sludge. Detergents are part of the additive package to clean-up existing deposits in the engine. They also disperse insoluble matter into the oil. Dispersants are also part of the additive package. Both the detergents and dispersants attach themselves to contaminated particles and hold them in suspension. The suspended particles are so finely divided that they can pass harmlessly between the mating surfaces and through the oil filter. This contamination is removed when the oil is changed.

(*vi*) **Cool Engine Parts:** About 60% of the cooling job of the engine is performed by the cooling system. It cools the upper part of the engine including the cylinder heads, cylinder walls, and valves. The crankshaft, the main and connecting rod bearings, the timing gears, the pistons and other components in the lower engine are cooled as the oil flows around the parts. The critical part is the continuous circulation of large quantities of oil.

6.12 DISPOSAL OF LUBRICANTS

Generally, lubricants are disposed directly in the environment as users discharge them into drains, onto the ground or directly into landfills. Approximately 40% of lubricants are released into the environment which causes significant damage to the environment. Also, the additives added to the lubricants are harmful to both flora and fauna. Further, used lubricants contain the oxidation products which are toxic. Common methods of lubricant disposal are discharge of used lubricant into water, landfills, lubricant burning, recycling of lubricant, use as weed-killer etc.

In most of the countries, there are strict regulations regarding discharge into water and landfill since even a small quantity of lubricant can cause severe pollution. In water, lubricating oil tends to settle down at the bottom, coating the substrate and whatever organisms live there. Oil film on water can also reduce the penetration of light into the water and, consequently, affect the rate of photosynthesis. When poured on the ground, lubricating oil can penetrate through the soil and can affect bacteria, plant, invertebrates and vertebrates. Lubricating oil clogs respiratory system of the organisms and make them unfit for human consumption.

Burning of lubricant as fuel, generally to produce electricity is also governed by regulations on account of the high level of additives present in them. Burning of lubricant produces number of pollutants. Thus, it must take place in specialized facilities that have special scrubbers to remove air-borne pollutants. The eco-friendly method of lubricant disposal is to recycle it. This method is quite economical. For example, used lubricating oil can be used as boiler fuel and good saving can be achieved.

6.13 ADDITIVES USED IN LUBRICANTS

Some of the additives added to improve properties of lubricants are as follows:

(*i*) **Viscosity Index Improvers:** The viscosity of multi-grade oils is controlled by viscosity-index improvers. These long-chain polymers function by uncoiling or dissociating at elevated temperature, which increases the oil's resistance to flow. However, at low temperatures, they are tight-balls, which insignificantly increase the oils resistance to flow.

(*ii*) **Detergents:** By neutralizing the acidity byproducts of lubricant oxidation and thermal decomposition, it may make the contaminants soluble as sludge of oxidized oil.

(*iii*) **Dispersants:** It controls contamination from low temperature operation. Both detergents and dispersants attach themselves to contaminant particles, and hold them in suspension. The contamination is removed from the engine when change of oil takes place.

(*iv*) **Oxidation Inhibitors:** Oxidation inhibitors perform the function of reducing oxygen attack on the lubricating base oil.

(*v*) **Corrosion and Rust Inhibitors:** The non-ferrous metals are easily protected by corrosion inhibitors coating which coats and form a barrier between parts and their environment. Rust Inhibitors protect iron/steel from oxygen attack as they form a protection screen over the metal surface.

(*vi*) **Anti-Wear Agents:** These anti-wear agents form a protective layer by chemical reaction with the metal surface.

(*vii*) **Extreme Pressure additive (antiseize or antiscuffing additive):** They also form a protective layer by chemical reaction with the metal surface. This increases the load at which scuffing or seizure occurs.

(*viii*) **Foam Depressants:** The foam depressants control the tendency of foam formation.

6.14 LUBRICANT DEGRADATION

There are various factors, which causes degradation of lubricant such as service time, operating temperature and contamination. Service time and operating temperature are directly related to each other. The useful life of a lubricant is extended when machine is operated at moderate operating temperatures while it is reduced when machine is operated at high operating temperatures.

Contamination of lubricant can result from various reasons such as poor maintenance of machine/equipment, poor handling techniques, system-ingested contaminants or system-generated contaminants etc. In IC engines, the ingested contamination is dirt, which results in wear of engine components. Such contamination is usually caused by a malfunction of the engine itself. Various types of system-generated contaminants may also be present such as antifreeze fluid, unburned fuel and products of combustion that can enter the lubricant crankcase through the piston ring area.

6.15 REPLENISHING MACHINE WITH FRESH LUBRICANT

Proper lubrication with timely replacement plays a significant role in maintaining the accuracy of the machine and enhances life of the machine. The process of lubrication, cleaning and replenishing machine with fresh

Lubricant should be carried out by trained person. The machine instruction's manual should be consulted for lubrication instructions. These manual have not only details of lubrication schedule but also specification of lubricant to be used.

6.15.1 Procedure of Changing Engine Oil

The general instructions of changing engine oil to be followed are as follows:

1. The vehicle should be raised and it should be ensured that vehicle is level and stable. Leakage should be checked visually.
2. Generally vehicles have drain plugs fitted at the place which is easily accessible. The position of the drain plug should be located and guard covers, if any, should be removed.
3. Correct wrench should be selected to loosen the plug. A large diameter pan should be positioned and plug should be removed.
4. After the lubricant is completely drained out, drain plug should be cleaned and put back with a new washer.
5. After checking instruction's manual, correct specification lubricating oil should be used in correct quantity.

6. Pump should be used to put fresh lubricating oil into the filler hole or dip-stick tube opening.
7. After each pump, a pause of few seconds should be given so that the lubricating oil flows to the bottom.
8. When the lubricating oil has reached to the bottom of the filler plug hole, or to the top of the mark on the dip-stick, pumping should be stopped.

6.15.1.2 Precautions

Following precautions should be followed while changing engine oil:

(*i*) Finger should not be inserted into the filler plug hole when engine is running.

(*ii*) If the vehicle has been driven just before to checking the oil level, the oil could be hot. Thus, when drain plug is removed, the hot lubricating oil may run out and result in burning. Hence, care must be taken in this regard.

(*iii*) Lubricating oil should be selected as per recommendations of the manufacturer. The instruction manual should also be consulted. Selection of wrong lubricating oil can result in machine failure.

(*iv*) A correct oil level is important. If the level of oil is too high, the oil may churn and aerate, causing bubbles that can trap heat and allow the oil to run hotter. This may cause early failure of bearings and gears. On the other hand, if the level of oil is too low, air may enter the system and heat generation can result in the burning of components.

SUMMARY

- The science which deals with reduction of friction by application of lubricant between the rubbing surfaces of bodies having relative motion is known as lubrication.
- Lubricant reduces the waste of energy, surface wear, damage, deformation and operational costs.
- Examples of solid lubricants are graphite, molybdenum disulfide, mica, chalk, talc, boron nitride, calcium fluoride, tungsten disulfide, etc.
- Examples of liquid lubricants are mineral oils, animal oils, vegetable oils and synthetic oils.
- A good lubricant should have high viscosity, high surface tension and good oiliness and adhesiveness.
- The measure of degree of fluidity of the oil and internal resistance offered against shear of the oil film is called viscosity.
- Oiliness is the property of the lubricant by virtue of which oil can stick on to the surface of the machine parts operating under high pressure.
- Adhesiveness is the ability of the lubricant to stick to the metal surface to obstruct metal to metal contact.
- The manufacturers usually recommend lubricants for their products in terms of specifications or required qualities.
- Greases are applied in those areas where a continuous supply of oil cannot be maintained, such as in open bearings or gears.
- Hydrodynamic lubrication is a system of lubrication in which the load supporting film is created by the shape and relative motion of the sliding surfaces.

- Hydrostatic Lubrication is a system of lubrication in which, the load supporting fluid, separating the two surfaces is generated by an external source, as the pump, supplying sufficient fluid under pressure.
- Dispersants in lubricant controls contamination from low temperature operation.
- Oxidation inhibitors perform the function of reducing oxygen attack on the lubricating base oil.
- Various factors, which cause degradation of lubricant, are service time, operating temperature and contamination.
- The useful life of a lubricant is extended when machine is operated at moderate operating temperatures while it is reduced when machine is operated at severe operating temperatures.
- Contamination of lubricant can result from various reasons such as poor maintenance of machine/ equipment, poor handling techniques, system-ingested contaminants or system-generated contaminants etc.

ASSIGNMENT

1. Student should be acquainted with lubricating devices, oil cans and grease gun.

EXERCISES

1. Explain the concept of lubrication. Why is it necessary in bearings and moving parts of a machine?
2. Discuss the purpose of lubrication.
3. What are the desired characteristics of lubricants?
4. Discuss briefly the important properties of lubricating oils which are useful for their evaluation.
5. What is a lubricator?
6. How lubricants are classified?
7. Differentiate between hydrodynamic and hydrostatic lubrication
8. With the help of neat sketch, discuss the working of a screw cap lubricator.
9. With neat sketch, explain the working of a tell-tale lubricator.
10. Explain with neat sketch working of drop feed oil lubricator.
11. What is a glass bottle lubricator? Explain it with a neat sketch.
12. What is a wick feed lubricator? Explain its working with neat sketch.
13. What is splash lubrication? Where it is used? Explain it with a neat sketch.
14. Sketch and explain the working of a ring oil lubricator.
15. Write notes on full pressure lubrication. Also, draw its neat sketch.
16. With neat sketch, describe working of oil gun.
17. Explain working of grease gun with neat sketch.
18. What are the functions of engine oil?
19. Discuss the methods of lubricant disposal.
20. Why are additives added to lubricants? Describe some of the additives.
21. What causes degradation of lubricant?
22. Discuss procedure of changing engine oil.

❑❑❑

7 CHAPTER

Inspection and Quality Control

7.1 INSPECTION

Inspection is a significant aspect of production control. The term inspection has been defined in number of ways by different authors. According to Dr. W.R. Speringle, "Inspection is the process of measuring the quality of a product or service in terms of established standards." Inspection is an important tool for controlling the quality of product. It is sort of post-mortem operation performed after completion of process. Products can be checked for quality at various stages. Defective products can be easily separated from non-defective ones, so that the defective products don't reach the customers.

The main objectives of inspection are:

(*i*) To collect information related to the product with established quality norms.

(*ii*) To find out defects in raw materials and faults in process that can cause damage to the product at later stages.

(*iii*) To separate and discard the defective or poor quality products. This will help the organization in quality control of products.

(*iv*) To establish or improve the reputation of the organization by protecting customer from receiving defective or poor quality products.

7.2 IMPORTANT DECISIONS FOR INSPECTION

Before starting inspection work, we must plan and take the following decisions regarding inspection:

(*i*) What to inspect?

(*ii*) When to inspect?

(*iii*) Where to inspect?

(*iv*) How to inspect?

(*v*) How much to inspect?

7.3 TYPES OF INSPECTION

Some of the important types of inspection are as follows:

(*i*) **Trial-run inspection:** In this type of inspection, tools, fixtures, jigs and gauges are inspected before start of production work.

(*ii*) **First piece inspection:** In this type of inspection, 3-4 products are inspected and if they are found to be satisfactory then it is assumed that products manufactured afterwards will also be satisfactory.

(*iii*) **Working inspection:** In this method, the products should be inspected while they are in process.

(*iv*) **Sample inspection:** In this type of inspection, certain percentage from batch of products is inspected and the entire batch is judged from the result obtained.

(*v*) **Operation inspection:** It is method, inspection is carried out at the completion of operation before the work in process is passed on for another operation.

(*vi*) **Final inspection:** In this process, products are inspected after manufacturing work is completed and before sending them to the store.

(*vii*) **Functional inspection:** In this method of inspection, products are checked for accuracy after completion of assembly.

(*viii*) **Floor inspection:** In this method, inspection work is carried out at or near the machine.

(*ix*) **Centralized inspection:** In this method, inspection work is done in the inspection rooms. The products to be inspected are taken to special rooms where precision measuring devices are utilized.

7.4 STAGES OF INSPECTION

Various stages of inspection are as given below:

(*i*) Inspection of incoming materials.

(*ii*) Inspection during work in progress.

(*iii*) Inspection of final product.

7.5 QUALITY

The term 'Quality' includes all the characteristics, which describe the acceptability of a product such as:

(*i*) Fitness for the purpose it has been manufactured.

(*ii*) To maintain the degree of conformance of all significant features to all the aspects of customer need limited by the cost.

(*iii*) Conformance of requirements

(*iv*) Quality attributes like its life, appearance, taste, smell, performance etc.

(*v*) The totality of features of a product that satisfy quality norms.

It can be fairly concluded that the quality is essential to complete effectively with international standards.

7.6 QUALITY CONTROL

Quality control may be defined as the process by means of which product of uniformly acceptable quality is manufactured. It is employed to ensure a product meet certain standard of quality. Also, products should be reliable, satisfactory and fiscally sound. Quality control involves routinely collection and analyzing followed by immediate corrective action.

Quality control includes inspection, testing and comparing the quality attributes of a product with standard specifications. Then deciding whether the product is acceptable or not. Thus, the objective of quality control program is to recognize products or processes which don't meet standards of quality. If such a problem is identified, the job of a quality control team is to stop production temporarily. Depending on the particular product or process, as well as the nature of problem identified, production may not cease entirely. Quality control is not just concerned with product or processes, but also employees. If an organization has employees that don't have adequate knowledge, skills or training, quality will decline for sure. Thus, quality control program also includes proper training of staff.

The following are some important objectives of quality control:

(*i*) To set up standards of quality that is acceptable to the customers limited by the cost.

(*ii*) To locate, identify and rectify faults so as to control defects, scrap and unnecessary wastage of material.

(*iii*) To take required steps so as to maintain the quality of the products.

(*iv*) To ensure that sub-quality products do not reach the customers.

(*v*) To set up quality norms to enable interchangeable manufacture in mass production.

Quality control in an organization provides following advantages:

(*i*) Quality of product is enhanced which can greatly affect sales.

(*ii*) Scrap and wastage of materials are minimized.

(*iii*) Reputation of organization improves.

(*iv*) The relationship between organization and customer are improved.

(v) Uniformity in quality of products is achieved.

(vi) Organization can face global competition.

7.7 FACTORS AFFECTING QUALITY

Apart from four M's (men, machines, materials and manufacturing conditions), there are some other factors which greatly affect the quality of product. These factors are:

(*i*) Customer's demand.

(*ii*) Capability of organization to invest.

(*iii*) Management policies and interest for maintaining quality.

(*iv*) Methods of production.

(*v*) Design of products.

Some other factors can severely damage organization's reputation regarding quality. These are poor methods of packing, transportation and after sales services.

7.8 DIFFERENCE BETWEEN INSPECTION AND QUALITY CONTROL

S. No.	*Inspection*	*Quality Control*
1.	Inspection is a process of separating acceptable and non-acceptable products from batch of products.	Quality Control is aimed at identifying causes of defects and rectifying them.
2.	Inspection is a part of the quality control program.	Quality control includes inspection alongwith corrective and preventive measures.
3.	Inspection increases the scrap and wastage.	Quality control reduces scraps and wastage.
4.	Inspection checks and passes the quality of final products.	Quality control develops quality in the product.

7.9 QUALITY ASSURANCE

Quality assurance is defined as "planned and systematic effort by a manufacturer to assure that its products conform to the existing quality standards". It emphasizes on the fact that the quality of the product is not just the responsibility of persons involved with quality control but also of every person associated in designing, production, testing and sales/ after sales service etc. Quality assurance provides confidence to the top management that all persons concerned are performing their job effectively with full competence and that the quality of product is as per laid down quality standards.

The important objectives of a quality assurance system are as follows:

(*i*) It plans, develops and establishes quality policies.

(*ii*) It conducts special studies and investigations related to quality.

(*iii*) It monitors quality of raw material.

(*iv*) It checks the accuracy of measuring instruments

(*v*) It evaluates and controls the quality of product in market.

(*vi*) It establishes feedback system.

(*vii*) It provides adequate training to the worker.

7.10 QUALITY CIRCLE

Quality circle is a small volunteer group of workers of the same work area doing similar work, who are trained to identify, analyze and solve work-related problems. This group presents their solutions to management in order to achieve improvement in the performance of the organization.

Quality circle meetings are important part of quality circle's working. These meetings should be attended by all the members of the quality circle group. Usually, meetings take place once a week. Quality circle group meets to discuss problems and then solve them. The solution is presented to the management for approval. The solution is implemented after getting approval from the management and is monitored by the quality circle group.

The objectives of Quality Circles are as follows:

(*i*) It develops problem solving capability.

(*ii*) It develops team spirit.

(*iii*) It eliminates inter-personal conflicts.

(*iv*) It develops positive working environment.

(*v*) It improves communication.

(*vi*) It reduces chances of error.

(*vii*) It promotes leadership qualities and personal development.

(*viii*) It enhances quality.

(*ix*) It enhances motivation of the worker.

(*x*) It develops harmonious relationship between manager and worker.

7.10.1 Benefits of Quality Circle

Following are the benefits of quality circle:

(*i*) It helps in self development in the workers.

(*ii*) It improves quality.

(*iii*) It improves productivity.

(*iv*) It promotes leadership qualities among participants.

(*v*) It develops team spirit.

(*vi*) It develops problem solving capacity.

(*vii*) It promotes improvement in products and services.

(*viii*) It increases worker's job satisfaction.

7.11 STATISTICAL QUALITY CONTROL

Statistical Quality Control (SQC) is the method of estimating the quality of the whole from the quality of samples. It utilizes the statistical tools for achieving it. Data are collected, analyzed and interpretations are made to achieve control in quality level. SQC involves analysis of the characteristics of a production output by interference from the sampling the output. The analysis, in turn, is used to take the necessary corrective measures that will keep the process under complete control. SQC is aimed to detect trends that will result in the defective production. The objective of SQC is to find out the causes for such trends and then to take necessary corrective action before defects are produced.

The main objectives of SQC are:

(*i*) It prevents the production of defective parts by inspection and establishes control over production process. It is not required to wait for inspection till completion of production, which is too late for corrective measure.

(*ii*) It reduces the inspection cost. It also ensures that quality of the product is controlled by adopting sample inspection techniques that utilizes statistical tools.

Statistical Quality Control (SQC) offers the following advantages:

(*i*) Better quality is achieved and maintained.

(*ii*) Uniformity in quality of products is achieved.

(*iii*) Inspection cost is reduced.

(*iv*) Utilization of equipment is more efficient.

(*v*) Scrap and wastages are minimized.

(*vi*) Raw materials are properly utilized.

(*vii*) Manufacturer-consumers relationships are improved.

7.12 SOME IMPORTANT TERMS

7.12.1 Variables

Measurement by variables incorporates the use of a continuous scale, in which the results of the quality can be expressed quantitatively. For example, dimensions, weight, temperature etc. Variables can be grouped into two categories.

(*i*) **Chance variables:** These variables are random and are difficult to be identified. Such variables are not easily avoidable. For example, variation in the properties of raw material, imperfection in the machine design and inherent inaccuracy of machine parts etc.

(*ii*) **Assignable variables:** These variables are random and can be identified. Such variables are easily avoidable. For example, improper sequence of manufacturing operations, worn out parts of machine, defective machine parts and variation in room temperature etc.

7.12.2 Attributes

Attributes are those measurements in which a quality characteristics is present or absent, or falls into a limited number of discrete categories. In other words, we say attributes are those quality characteristics, which are not measured by gauge to determine whether these are acceptable or not acceptable. For example GO and NO GO gauges measurement.

7.13 CONTROL LIMITS

The variation observed in the measured quantities can be represented by a frequency distribution. Depending upon the situation, it can have various shapes. The study of the frequency distribution is of great interest, when quality characteristics are influenced only by chance variable. The process under influence of chance variable is said to be in a state of control. Under such situation, the frequency distribution will be represented by a bell-shaped symmetrical curve known as normal distribution curve, as shown in Figure 7.1.

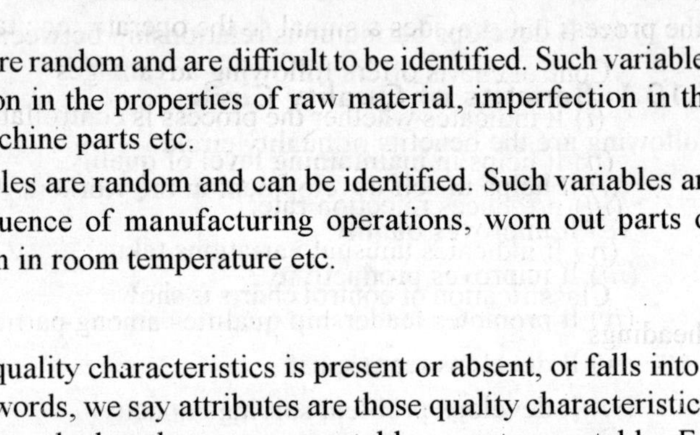

Figure 7.1 Normal Distribution Curve

It can be seen from the curve that the frequency of variation of a number of observations for the various values around the central value (arithmetic mean). It represents a population of infinite size. It is asymptotic about *X*-axis. It is described fully in terms of $\bar{X}$ and σ. The normal distribution curve has following important properties.

(*a*) $\bar{X} \pm \sigma$ occupy 68.2% of observation

(*b*) $\bar{X} \pm 2\sigma$ occupy 95.4% of observation

(*c*) $\bar{X} \pm 3\sigma$ occupy 99.7% of observation

7.14 CONTROL CHART

Control chart can be defined as a chronological graphical representation of actual product-quality characteristics with limits reflecting the ability to produce as shown by past experience on the product characteristics. Such charts are based upon statistical sampling theory according to which an adequately sized sample drawn at random from a batch represents the batch. These charts themselves can't control the process but provides a signal so the operator can take necessary corrective action.

Control charts offers following advantages:

(*i*) It indicates whether the process is controllable or uncontrollable at particular instant of time.
(*ii*) It helps in maintaining level of quality.
(*iii*) It reduces rejection rate.
(*iv*) It indicates unusual variations taking place.

Classification of control charts is shown in Figure 7.2. These charts are discussed in the following headings.

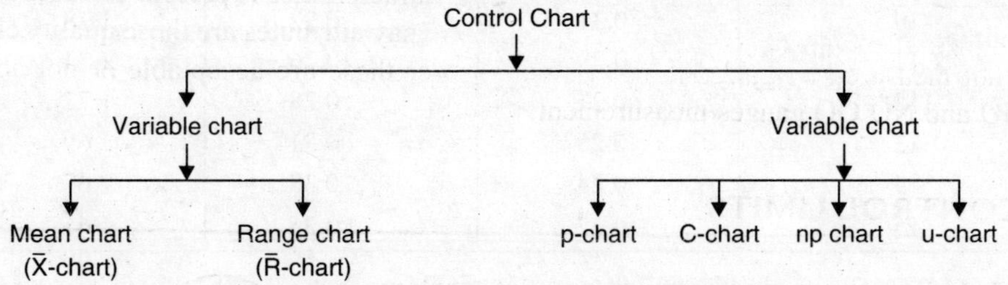

Figure 7.2 Classification of control charts

7.14.1 Variable Chart

In variable charts, the method of inspection is by variables. The $\bar{X}$ and $\bar{R}$ charts are examples of variable charts.

$\bar{X}$ and $\bar{R}$ charts: A number of samples of components are taken over a period of time. Each sample is taken at random and the size of sample is kept from 5 to 15 components. For each sample, the average value (X) of all measurements and the range (R) are determined. The grand average $\bar{X}$ (average value of all the X) and $\bar{R}$ (average value of all the ranges) is determined. From these, we find out the control limits for the $\bar{X}$ and $\bar{R}$ -charts.

For $\bar{X}$ charts,

Upper control limit, $UCL_{\bar{X}} = \bar{X} + A_2 \bar{R}$

Lower control limit, $LCL_{\bar{X}} = \bar{X} - A_2 \bar{R}$

For $\bar{R}$ charts,

Upper control limit, $UCL_{\bar{R}} = D_4 \bar{R}$

Lower control limit, $LCL_{\bar{R}} = D_2 \bar{R}$

The factors A_2, D_4 and D_3 are known as the statistical factors depend on the number of components per sample. The value of these factors is obtained from Statistical Quality Control Table. For reference, these are given in Table 7.1.

Table 7.1 Statistical Quality Control Table

No. of components in a sample	A_2	D_3	D_4
2	1.88	0	3.27
3	1.02	0	2.57
4	0.73	0	2.28
5	0.58	0	2.11
6	0.48	0	2.00
7	0.42	0.08	1.92
8	0.37	0.14	1.86
9	0.33	0.18	1.82
10	0.31	0.22	1.78
11	0.29	0.26	1.74
12	0.27	0.28	1.72
13	0.25	0.31	1.69
14	0.24	0.33	1.67
15	0.22	0.35	1.65

Example 7.1. *Following observations were made from the samples of 5 items each.*

Sample No.	*1*	*2*	*3*	*4*	*5*	*6*	*7*	*8*	*9*	*10*
X	*7.5*	*7.0*	*9.5*	*8.0*	*10.0*	*11.5*	*4.0*	*3.5*	*10.5*	*4.5*
R	*3*	*2*	*3*	*4*	*3*	*3*	*2*	*2*	*2*	*2*

Determine upper and lower control limits for $\bar{X}$ and $\bar{R}$ chart.

Also, draw the $\bar{X}$ and $\bar{R}$ chart.

Solution:

$$\text{Grand average, } \bar{X} = \frac{\text{Sum of all } X}{\text{Total no. of samples}} = \frac{\Sigma X}{10} = \frac{76}{10} = 7.6$$

$$\text{Average of all ranges, } \bar{R} = \frac{\text{Sum of all } R}{\text{Total no. of samples}} = \frac{\Sigma R}{10} = \frac{26}{10} = 2.6$$

From Table 7.1, the values of A_2, D_4 and D_3 can be noted for sample size $(n) = 5$,

$$A_2 = 0.58$$
$$D_3 = 0$$
$$D_4 = 211$$

For $\overline{X}$ -chart,

Upper control limit, $UCL_{\overline{X}} = \overline{X} + A_2\ \overline{R} = 7.6 + (0.58 \times 2.6) = 9.11$

Lower control limit, $LCL_{\overline{X}} = \overline{X} - A_2\ \overline{R} = 7.6 - (0.58 \times 2.6) = 6.09$

For $\overline{R}$ -chart,

Upper control limit, $UCL_{\overline{R}} = D_4\ \overline{R} = 2.11 \times 2.6 = 5.49$

Lower control limit, $LCL_{\overline{R}} = D_3\ \overline{R} = 0 \times 2.6 = 0$

Now, charts for $\overline{X}$ and $\overline{R}$ are plotted as shown in Figure 7.3 (*a*) and 7.3 (*b*) taking abscissa as sample no. and ordinate as *X* and *R* respectively.

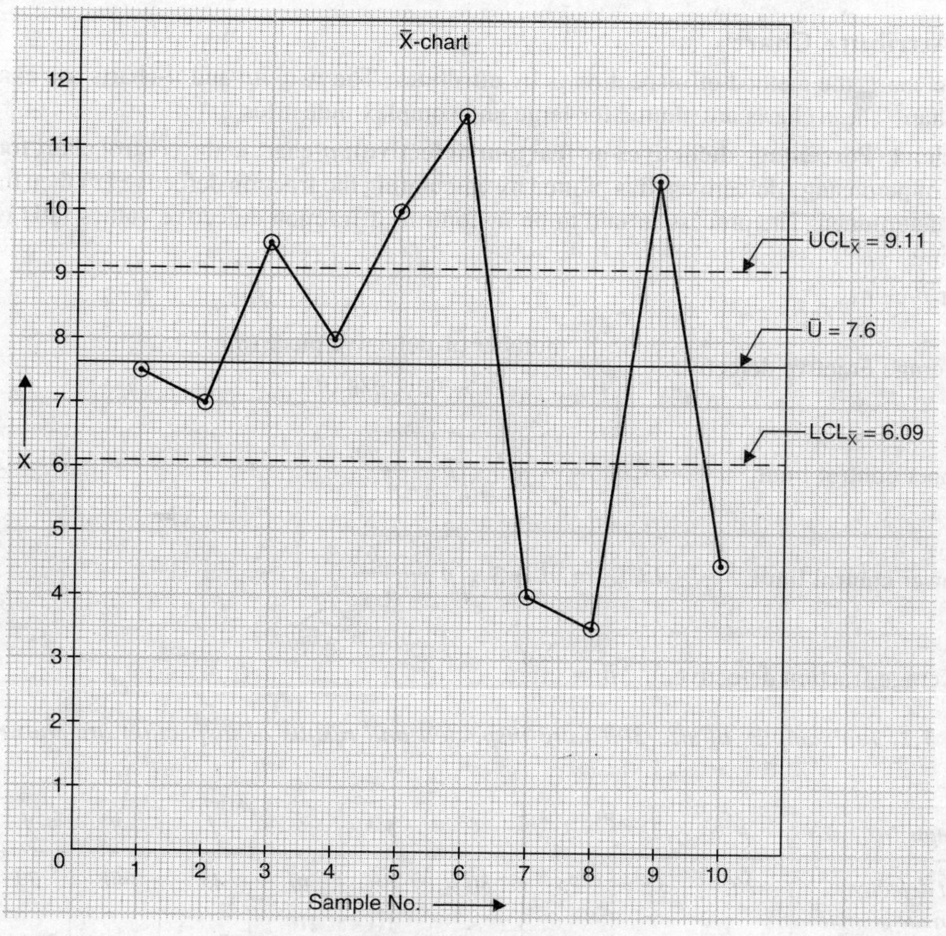

Figure 7.3 (a)

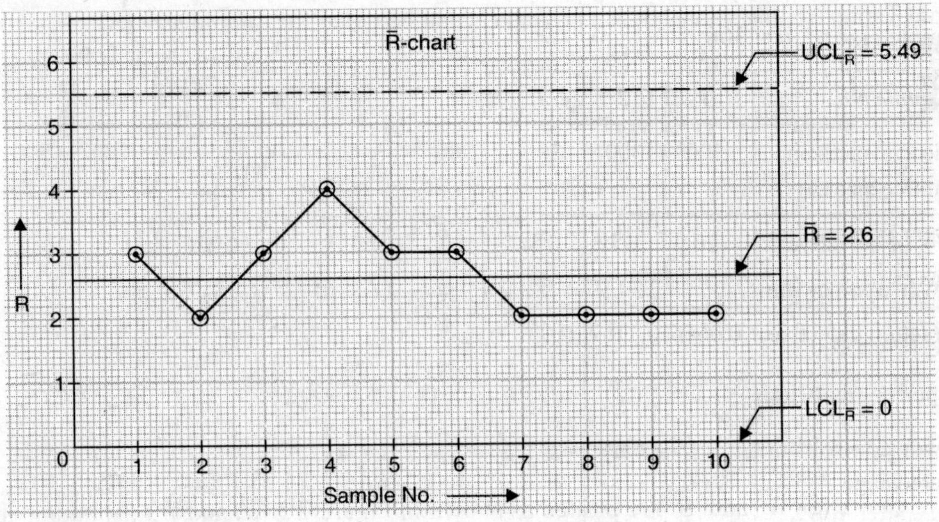

Figure 7.3 (b)

7.14.2 Attribute Chart

In attribute chart, the method of inspection is by attributes. The *p, c, np* and *u*-charts are examples of attribute chart. These charts are plotted for large and complex industries.

***p*-chart (Percentage defectives or fraction defective):** *p*-chart is the control chart for fraction defectives of percentage defectives. It is plotted by calculating the fraction defective (p) then the control limits are determined. The process is said to be in control, if fraction defective values falls within the control limits.

For *p*-chart,

Average Fractive defective, $$p = \frac{\text{number of defective piece } (d)}{\text{sample size } (n)}$$

Upper control limit, $$UCL_{\bar{p}} = \bar{p} + 3\sqrt{\bar{p}\,\frac{(1-\bar{p})}{n}}$$

Lower control limit, $$LCL_{\bar{p}} = \bar{p} + 3\sqrt{\bar{p}\,\frac{(1-\bar{p})}{n}}$$

Average fraction defective, $$\bar{p} = \frac{\Sigma d}{n}$$

Example 7.2. *Ten samples of size 300 were inspected and results of inspection are shown in table below:*

Sample No.	*1*	*2*	*3*	*4*	*5*	*6*	*7*	*8*	*9*	*10*
No. of defective pieces in the sample	20	35	25	30	20	40	45	40	20	35

Draw the chart for fraction defectives.

Solution:

Sample size, $n = 300$

S.No.	Sample No.	1	2	3	4	5	6	7	8	9	10
1	Piece inspection sample size (n)	300	300	300	300	300	300	300	300	300	300
2	No. of defective pieces (d)	20	35	25	30	20	40	45	40	20	35
3	Fraction defective, $p = \frac{d}{n}$	0.0666	0.1167	0.083	0.100	0.0666	0.1333	0.150	0.1333	0.0666	0.1167
4	Percentage defective, 100 p	6.66	11.6	8.3	10	6.66	13.33	15.0	13.33	6.66	11.67

$$\text{Average fraction defective } (\bar{p}) = \frac{\Sigma d}{n} = \frac{310}{10 \times 300} = 0.1033$$

Upper control limit, $$UCL_{\bar{p}} = \bar{p} + 3\sqrt{\bar{p}\frac{(1-\bar{p})}{n}}$$

$$= 0.1033 + 3.\sqrt{\frac{0.1033\,(1-0.1033)}{300}}$$

$$= 0.1033 + 3.\sqrt{\frac{0.1033 \times 0.8967}{300}}$$

$$= 0.1033 + 3 \times \sqrt{0.000309}$$
$$= 0.1033 + 3 \times 0.0176 = 0.1561$$

Lower control limit, $$LCL_{\bar{p}} = \bar{p} - 3\sqrt{\bar{p}\frac{(1-\bar{p})}{n}}$$

$$= 0.1033 - 3.\sqrt{\frac{0.1033\,(1-0.1033)}{300}}$$

$$= 0.1033 - 3.\sqrt{\frac{0.1033 \times 0.8967}{300}}$$

$$= 0.1033 - 3 \times \sqrt{0.000309}$$
$$= 0.1033 - 3 \times 0.0176 = 0.1033 - 0.0528 = 0.0505$$

Now, *p*-chart is plotted as shown in Figure 7.4 taking abscissa as sample no. and ordinate as *p*.

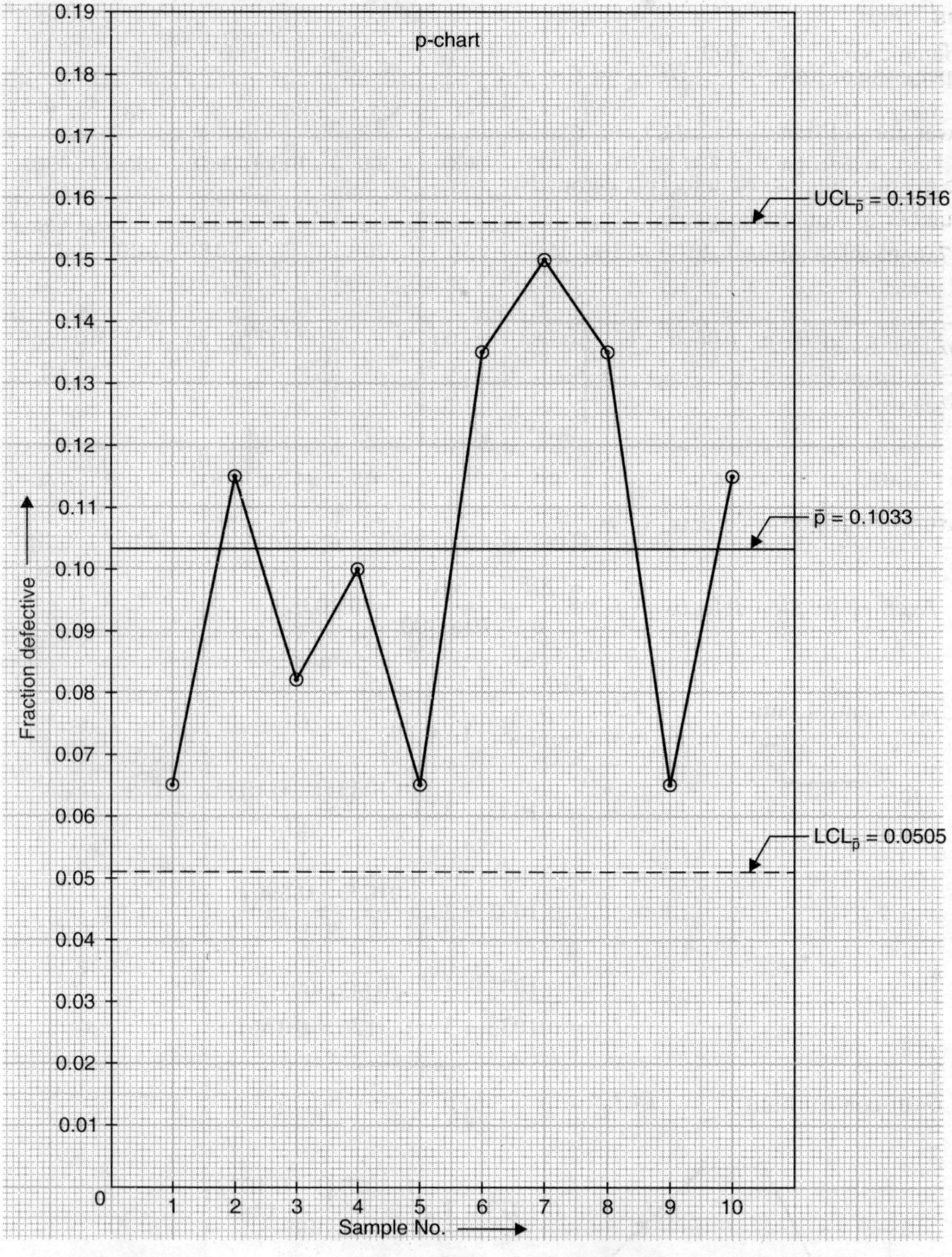

Figure 7.4

***C*-chart (Number of defects per unit):** In *C*-chart, number of defects in a sample or defects in a piece are considered. The construction of *C*-chart is similar to *p*-chart but the control limits are calculated on poisson's distribution. *C*-charts are used for large and complex parts.

For C-chart,

Upper control limit, $UCL_{\bar{C}} = \bar{C} + 3\sqrt{\bar{C}}$

Lower control limit, $LCL_{\bar{C}} = \bar{C} - 3\sqrt{\bar{C}}$

The average number of defects is detoned by $\bar{C}$ and is the ratio of the total number of defects to the total number of samples inspected.

Example 7.3. *Plot the control chart for the number of defects for the following inspection results of castings.*

No. of castings	*1*	*2*	*3*	*4*	*5*	*6*	*7*	*8*
No. of defects	*2*	*4*	*4*	*3*	*2*	*1*	*2*	*6*

Solution:

$$\text{Average number of defects, } \bar{C}, = \frac{\text{Total no. of defects}}{\text{Total no. of samples inspected}}$$

$$= \frac{2+4+4+3+2+1+2+6}{8}$$

$$\bar{C} = 24/8 = 3$$

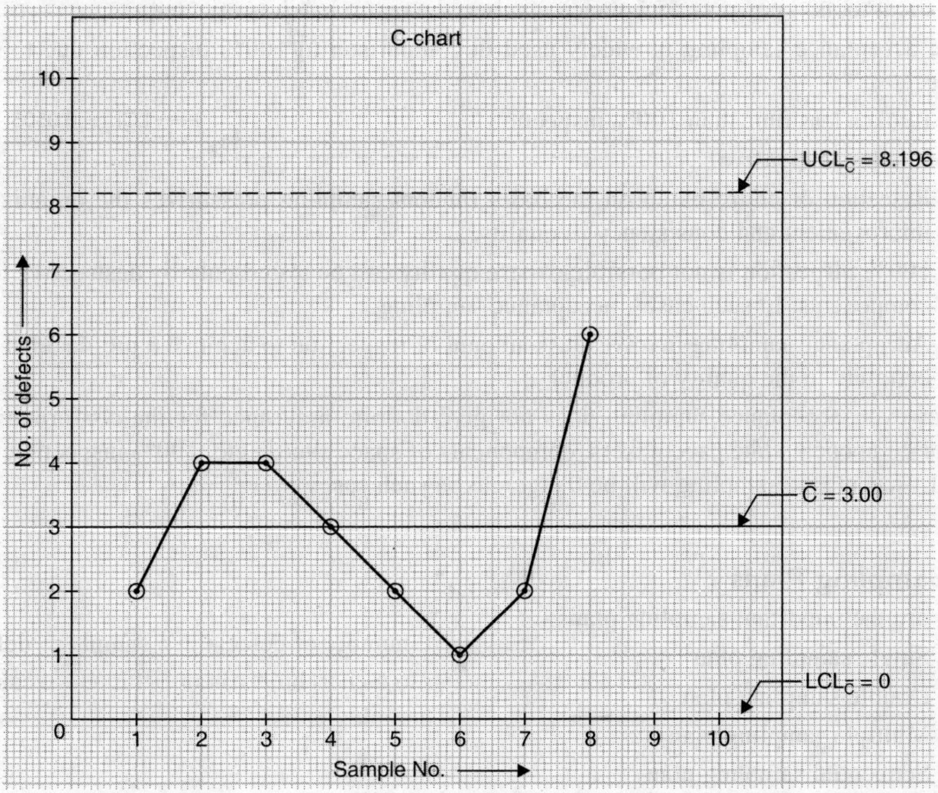

Figure 7.5

Upper control limit, $UCL_{\bar{C}} = \bar{C} + 3\sqrt{\bar{C}} = 3 + 3\sqrt{3}$

$= 3 + 3 \times 1.732 = 3 + 5.196 = 8.196$

Lower control limit, $LCL_{\bar{C}} = \bar{C} - 3\sqrt{\bar{C}}$

$= 3 - 3\sqrt{3} = 3 - 3 \times 1.732$

$= 3 - 5.196 = -2.196 = 0$

As there cannot be negative defects, $LCL_{\bar{C}}$ is taken as zero.

Now, C-chart is plotted as shown is Figure 7.5 taking abscissa as sample no. ordinate as no. of defects.

7.15 ACCEPTANCE SAMPLING

Acceptance sampling is a quality management technique used to distinguish good batch of items from bad batch. A bad batch is one which contains too many defective items.

Acceptance sampling is an act of drawing out some random samples from a batch and inspecting them to decide whether the entire batch is acceptable or not. Thus, the objective of acceptance sampling is to decide the acceptability of the batch. It only provides a technique of sentencing batches, *i.e.* calling them good or bad. It does not control the quality during manufacturing stages. A sample from the batch is drawn and inspected. If the number of defects is more than the stated number (the number obtained using statistical tools) known as acceptance number, the entire batch is rejected.

The acceptance sampling is widely used in comparison to 100% inspection due to following advantages:

(*i*) It takes less time than 100% inspection.

(*ii*) It is less expensive.

(*iii*) Rejection of entire batch on the basis random sampling pressurizes the organization, for improvement and consistency in quality.

(*iv*) 100% inspection is not possible, where quality can be tested only by destructive testing the components. In such cases, acceptance sampling.

(*v*) Acceptance sampling involves less amount of inspection to achieve a pre-decided degree of certainty in the quality of product.

However, acceptance sampling does not give definite assurance for the conformation of the quality. For components like pacemakers, components of space shuttle etc, 100% inspection should be carried out. Acceptance sampling will be dangerous in such cases.

7.15.1 Sampling Plans

Following are the important sampling plans:

(*i*) Single sampling plan

(*ii*) Double sampling plan

(*iii*) Multiple sampling plan

(*iv*) Sequential sampling plan

(*i*) **Single sampling plan:** One sample of items is selected at random from the batch and the disposition of the batch is determined from the resulting information. This plan is generally denoted as (*n*, *c*) plan for a sample size n and acceptance number *c*. The entire batch is rejected, if the number of defects is more than acceptance number. This type of plan is easiest plan and commonly employed although it is not much efficient.

(*ii*) **Double sampling plan:** In this type of sampling plan, the decision of acceptance and rejection of the batch is based on two sample sizes. After the first sample is tested, there are three possibilities:

(*i*) Accept the batch

(*ii*) Reject the batch

(*iii*) No decision

If the outcome is no decision, and a second sample is taken. Then the results of both the samples are combined and final decision is taken in the basis of this information.

(*iii*) **Multiple sampling plan:** Multiple sampling plan is the extension of the double sampling plan. In this type of sampling plan, more than two samples are required to reach a conclusion. The advantage of multiple sampling is smaller sample sizes.

(*iv*) **Sequential sampling plans:** This type of sampling plan is the extension of multiple sampling plan. In this plan, items are selected from the batch one at a time and after inspection of each item a decision is made to accept or reject the batch or select another unit.

7.16 BASIC CONCEPT OF TOTAL QUALITY MANAGEMENT (TQM)

Concept of total quality management originated in the 1950's and gained popularity since the early 1980's. Now, it is recognized as a generic management tool. It gives description of the work culture, attitude and organization of a company that strives to provide customers with products that fully satisfy their requirements.

Total Quality Management (TQM) is a management philosophy for continuously improvement of the quality of the products. TQM functions on the premise that the quality of products and processes is the responsibility of everyone who is involved with the creation or consumption of the products. TQM emphasizes on the involvement of management, workers, suppliers and customers, in order to meet customer needs and fulfill organizational objectives.

TQM is the foundation for activities, which include:

(*i*) Commitment by management and workers

(*ii*) Meeting requirements of customer

(*iii*) Morale of the worker

(*iv*) Quality control

(*v*) Quality system improvement

(*vi*) Reducing cost of the product

(*vii*) Quality circle

(*viii*) Improvement teams

(*ix*) Challenging quantified goals

(*x*) Management-customer relationship

(*xi*) Focus on processes improvement plans

(*xii*) Leadership
(*xiii*) Supplier quality management
(*xiv*) Vision and plan statement
(*xv*) Evaluation
(*xvi*) Product design
(*xvii*) Recognition and reward
(*xviii*) Education and training

7.17 AN INTRODUCTION TO ISO 9000

The important of standardization has been realized throughout the world and hence, International standardization has gained wide popularity. The objective of International Organization for Standardization (ISO) is to provide the development of standardization and the related activities through the globe. The ISO series program provides standards for documentation and audits as part of a quality management system.

With increased attention on quality issues, the ISO brought out 6 standards- One of standard terminology (ISO 8402) and 5 standards (known as ISO 9000 series). ISO 9000 are the internationally accepted standard for quality management system for designing, manufacturing and servicing the products. This quality system is audited by external auditor. ISO 9000 series can be implemented by an organization to ensure that their quality assurance system is effective. It is applicable to all types of organization and industries. Conformance to ISO 9000 guarantees that an organization delivers quality products/ services.

ISO 9000 series is comprise of five parts- ISO 9000, ISO 9001, ISO 9002, ISO 9003 and ISO 9004. Bureau of Indian Standards (BIS) is the Indian representative to ISO. Realizing the significance of ISO 9000 series, Bureau of Indian Standards has adopted these ISO standards and brought them out as IS 14000 series of Indian Standards. The corresponding Indian Standards are listed below.

(*i*) *ISO 8402/ IS 13999*: Quality system deals with vocabulary used
(*ii*) *ISO 9000/ IS 14000*: Quality systems- Guidelines for selection and use of standards on quality systems.
(*iii*) *ISO 9001/ IS 14001*: Quality systems-Model for quality assurance in design, development, production, installation and servicing.
(*iv*) *ISO 9002/ IS 14002*: Quality systems-Model for quality assurance in production and installation.
(*v*) *ISO 9003/ IS 14003*: Quality systems-Model for quality assurance in final inspection and test.
(*vi*) *ISO 9004/ IS 14004*: Quality Management and Quality system elements-Guidelines.

7.17.1 Advantages of ISO 9000 Series

Following are some of the advantages of implementing ISO 9000 series.

(*i*) It develops better quality in product and service.
(*ii*) It provides competitive edge in national and international market.
(*iii*) It fulfils requirements of the customers and brings confidence to the customers.
(*iv*) It provides consistency in output.

(v) It reduces wastes and time consuming rework.

(vi) It reduces complaints by the customers.

(vii) It results in harmonization of standards.

SUMMARY

- In words of Dr. W.R. Speringle, "Inspection is the process of measuring the quality of a product or service in terms of established standards."
- In first piece inspection, 3-4 products are inspected and if they are found to be satisfactory then it is assumed that products manufactured afterwards will also be satisfactory.
- In sample inspection, certain percentage from batch of products is inspected and the entire batch is judged from the result obtained.
- Floor inspection is carried out at or near the machine.
- The term 'Quality' includes all the characteristics, which describe the acceptability of a product.
- Quality control can be defined as the process by means of which product of uniformly acceptable quality is manufactured.
- Quality control includes inspection, testing and comparing the quality attributes of a product with standard specifications.
- Quality assurance can be defined as "planned and systematic effort by a manufacturer to assure that its products conform to the existing quality standards".
- Quality circle is a small volunteer group of workers of the same work area doing similar work, who are trained to identify, analyze and solve work-related problems. This group presents their solutions to management in order to achieve improvement in the performance of the organization.
- Statistical Quality Control (SQC) is the method of estimating the quality of the whole from the quality of samples. It utilizes the statistical tools for achieving it.
- Measurement by variables incorporates the use of a continuous scale, in which the results of the quality can be expressed quantitatively.
- Attributes are those quality characteristics, which are not measured by gauge to determine whether these are acceptable or not acceptable.
- Control charts themselves can't control the process but provides a signal so that the operator can take necessary corrective measure.
- Acceptance sampling is an act of drawing out some random samples from a batch and inspecting them to decide whether the entire batch is acceptable or not.
- Total Quality Management (TQM) is a management philosophy for continuously improvement of the quality of the products.
- TQM functions on the premise that the quality of products and processes is the responsibility of everyone who is involved with the creation or consumption of the products.
- The ISO series program provides standards for documentation and audits as part of a quality management system.

ASSIGNMENT

1. **Students should be asked to visit a local manufacturing company and study the procedure used by them to control quality of their product.**

EXERCISES

1. Define inspection and give its objectives.
2. What are the main objectives of inspection?
3. Explain various types of inspection with suitable example.
4. Differentiate between trial-run and first piece inspection.
5. What is meant by working inspection?
6. What do you understand by sample inspection?
7. Define:
   (*i*) Final inspection (*ii*) Functional inspection
   (*iii*) Floor inspection (*iv*) Centralized inspection
8. What are the different stages of inspection?
9. What important decisions are to be made before starting inspection?
10. Define the term "quality".
11. What do you understand by quality control?
12. What are the chief objectives of quality control?
13. What are the advantages of quality control in an organization?
14. What are the different factors which affect the quality of product?
15. Differentiate between inspection and quality control.
16. What do you understand by quality assurance? Also, give important objectives of a quality assurance system.
17. What is meant by quality circle? What are the objectives of quality circles?
18. What are the benefits of quality circle?
19. Explain briefly concept of TQM.
20. Write short notes on ISO 9000.

❑❑❑

8

CHAPTER

Reliability and Maintainability

8.1 INTRODUCTION TO RELIABILITY ENGINEERING

Reliability Engineering is concerned with the application of reliability concept in industries. It covers life data analysis of products like equipments, components, systems and processes etc. Reliability Engineering includes all aspects of a product's life from its conception, subsequent design, manufacturing processes, its practical life, maintenance and availability. In short, it covers all three aspects of RAM, *i.e.,* Reliability, Availability and Maintainability.

8.2 BASIC CONCEPT OF RELIABILITY

Reliability refers to the fact that a product function without failure for desired periods during their whole design life from conception to junking. It emphasizes on the capability of a product to perform its desired function. Mathematically speaking, if one supposes that a product is performing its desired function at time equal to zero (*i.e.*, at the beginning of the time interval), Reliability thus is referred as the probability that the product will continue to perform its desired function without failure till a specific time under stated conditions like in specified environments and with desired confidence that can be specified, designed, predicted, tested and demonstrated. The meaning of product, used here, can be mechanical or electronic product, component, hardware, software program, manufacturing process etc.

Reliability is different from quality control, which implies the performance of a product till its entire lifetime while quality control is concerned with the performance of a product at one point of time, especially during the manufacturing stages. The reliability of the product is expressed in terms of probability (0 to 100%) as a function of time. Thus, the reliability of a product may be expressed as 98% which means working for two years.

8.3 SYSTEM RELIABILITY

8.3.1 Series System

If components are in series and each has a reliability R_1. If one component fails, the system fails.

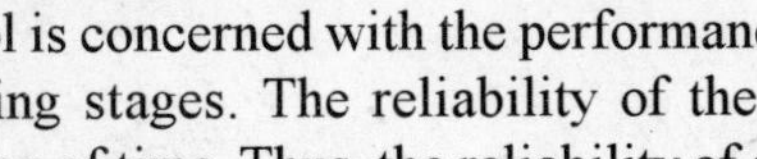

Figure 8.1 Series Reliability

The overall reliability of a series system is given as

$$R_{AB} = R_1 \times R_2 \times R_3$$
$$\text{If } R_1 = R_2 = R_3 = 0.90$$
$$R_{AB} = R_1 \times R_2 \times R_3 = 0.90 \times 0.90 \times 0.90 = 0.729$$

R_{AB} is always less than R_1 or R_2 or R_3.

8.3.2 Parallel System

If components are in parallel and each has a reliability R_1. If one component fails, the system does not fail.

$$R_{AB} = 1 - \text{probability (1 and 2 both fail)}$$

The probability of 1 failing is = $(1 - R_1)$

The probability of 2 failing is = $(1 - R_2)$

Then, overall reliability is $R_{AB} = 1 - (1 - R_1)(1 - R_2)$

If $R_1 = 0.9$ and $R_2 = 0.8$

$$R_{AB} = 1 - (1 - 0.9)(1 - 0.8) = 0.98$$

R_{AB} is always more than R_1 or R_2.

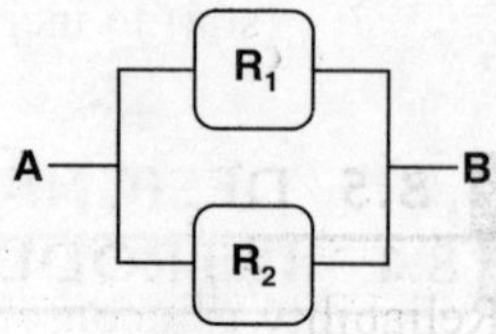

Figure 8.2 Parallel Reliability

8.4 RELIABILITY ANALYSIS

The Bath Tub Curve (as shown in Figure 8.3), is a generalized curve for the failure rates of product over operating time. This curve is important in minimizing the early failure period by use of quality control.

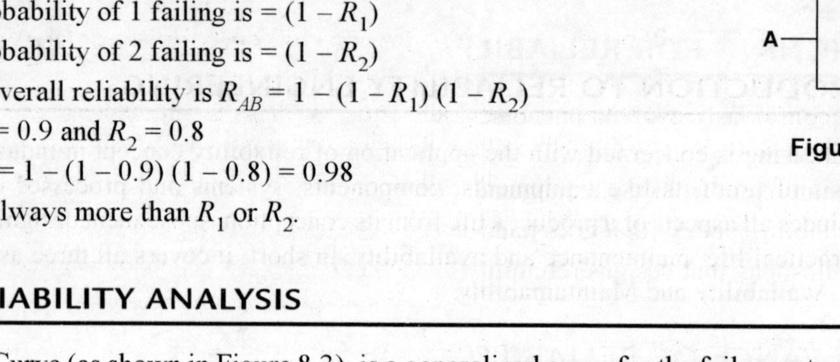

Figure 8.3 Bath Tub Curve

From inception to junking (wear out), most products go through three distinct phases. Figure 8.3 shows a typical life cycle curve in which the failure rate is shown as function of operating time. Failure rate mean the ratio of the number of failures during a particular interval to the average population during that interval.

(*i*) **Phase-I (Infancy or Burn-in-period):** Some products fail as soon as they are put in service. Poor design or manufacturing problems may be the main cause of failure in this period. This may result in calling a product to be legitimately bad. This is called infant mortality failures. The failure rate within this period (known as infant mortality rate) is significantly high. Sound manufacturer performs 'burn in' operation by testing the product for many days to so as to remove the problems so that the customer doesn't suffer due to them.

(*ii*) **Phase-II (Chance failure or Normal Operating Life):** If a product does not fail during Phase-I, it may continue to be trouble-free during its operating life. The failure rate during this phase is quite low. This phase represents the useful life of the product.

(*iii*) **Phase III (Wear out or Ageing):** After the product reaches a certain age, it enters the period where it begins to wear out, and failures start to increase rapidly. The period where failures start to increase is called the 'wear out' phase of the product life.

8.5 DESIGNING FOR RELIABILITY

Reliability is a consciously evolved phenomenon. It chiefly requires three key elements which are as follows:

(*i*) A commitment from top management ensuring reliability.
(*ii*) A reliability policy (that goes hand-in-hand with a quality control).
(*iii*) A philosophy that designs reliability at an early stage.

8.6 RELEVANCE OF RELIABILITY

In contemporary competitive and techno-savvy environment, reliability is an important product attribute, which involves:

(*i*) The industry's reputation.
(*ii*) Satisfaction of customer.
(*iii*) When a product fails to perform its function during the warranty period, the replacement and repair costs will negatively affect profits.
(*iv*) Improved reliability has a positive impact on future business which may lead to repeat business.
(*v*) The life cycle cost analysis may prove that the initial cost of any product might be higher, its overall lifetime cost shall be lower than their competitor's as their products needs less repairs and maintenance.
(*vi*) In developed/developing countries, customers expect their suppliers to have a remarkable reliability program.
(*vii*) Many companies publish their predicted reliability number to gain an edge over their competitors in the market who either do not like to publish their number or have lower numbers etc.

8.7 RELIABILITY IMPROVEMENT

By adopting following ways, the reliability of a system may be improved:

(*i*) Improvement in the quality of equipments.
(*ii*) Preference should be given to simplest possible designs.
(*iii*) The layout of equipment be improvised and innovated upon to facilitate.

(*iv*) Components having undergone reliability component testing can be used again.

(*v*) Proven designs and manufacturing methods may be used.

8.8 ADVANTAGES OF RELIABILITY

Reliability is advantageous in the following manner:

(*i*) It is of great help in good maintenance frequencies.

(*ii*) It gives good yield in proper replacement plans.

(*iii*) In a particular project, it may ensure better success evaluation.

(*iv*) Equipment availability is dependent on reliability.

(*v*) It is helps in making equipments for maximum safety.

(*vi*) Reliability is the hall mark of cost of produced good.

8.9 AVAILABILITY

Availability is defined as the probability that a product when used under stated conditions, without consideration for any maintenance, shall operate satisfactorily at any given time. Thus, we can say that availability is the probability that the system is properly operating when it is needed for use. In other words, It can be defined as the probability that a system is not failed or undergoing any repair when it is required to be used. Availability is always related to time, much like reliability.

Availability informs us about how to use time.

Availability is denoted by A,

Availability, A = Uptime/(Downtime + Uptime)

The unit of time is hours.

From design point of view,

Availability (Intrinsic), A_i = MTBF/(MTBF + MTTR)

Where,

MTBF = Mean time between failures

MTTR = Mean time to repair

Operational availability is defined as,

Availability (Operational), A_o = MTBM/(MTBM + MDT)

where,

MTBM = Mean time between maintenance

MDT = Mean down time.

8.9.1 Classification of Availability

Availability is the probability that the system has not failed or undergoing a repair action when it is needed to be used. It is classified as follows:

(*i*) **Point availability:** Point availability is defined as the probability that the system (or product) will be operational at any random time. It is also called as instantaneous availability. This is very much similar to the reliability function as it gives a probability that a system will function at the given time. At any given time t, the system will be operational, if the following conditions are fulfilled:

(*a*) It functioned properly during time t, with probability $R(t)$.

(*b*) It functioned properly since the last repair at time u, $0 < u < t$, with probability:

$$\int_0^t R(t-u)\, m(u)\, du$$

where, $m(u)$ being the removal density function of the system. The point availability is the sum of the above two probabilities:

$$A(t) = R(t) + \int_0^t R(t-u)\, m(u)\, du$$

(*ii*) **Mean availability:** Mean availability is defined as the proportion of time during a time period when the system is available for use. It represents the mean value of the point availability function over the period (o, t):

$$\overline{A(t)} = \frac{1}{t}\int_0^t A(u)\, du$$

(*iii*) **Steady state availability:** Steady state availability is defined as the limit of the point availability as time approaches infinity.

$$A(\infty) = \lim_{t \to \infty} A(t)$$

(*iv*) **Operational availability:** Operational availability is defined as the measure of availability that includes all experienced sources of down time, logistic downtime, etc. The expression for operational availability is:

$$A_o = \frac{\text{Uptime}}{\text{Operating Cycle}}$$

where, the operating cycle is the overall time period of operation under consideration and uptime is the total time in which the system was functioning during the operating cycle.

8.10 MAINTAINABILITY

If a system or product fails, it is possible to put it on an operating condition as soon as possible, this is known as maintainability. It is an inherent characteristic of the system design pertaining to ease, accuracy, safety and economy in performing maintenance functions. In short maintainability is the ease in time or resources with which a system or product can be modified to correct faults, bring performance to peak or adapt to a changed environment.

It may also be defined as the measures taken during the design, development and installation of a manufactured product that reduce required maintenance, man-hours, tools, logistic costs, skill levels and facilities etc. It can be measured in terms of Mean time to repair (MTTR) of a product, *i.e.*, the time required to perform each anticipated repair multiplied by the relative frequency with which that task is performed.

8.10.1 Advantages of Maintainability

Major advantages of maintainability are as follows:

(*i*) The product available to the users is reliable.

(*ii*) The product will face lesser defects during period of service.

(*iii*) It needs minimum maintenance effort in terms of man power, tools and cost-effectiveness.

(*iv*) It is safe in operation and handling.

8.10.2 Disadvantages of Maintainability

Maintainability has the following disadvantages:

(*i*) Maintainability does not improve performance.

(*ii*) Maintainability is usually introduced at design stage. It can't be put effectively at a later stage, particularly in a complex machine.

8.11 FAILURE

Generally speaking, a defect means fault, mistake, flaw, deficiency, or shortcoming of any type or situation. Failure means lack of success, deterioration, decay or stoppage etc. of an item for any reasons. Industrial failures may be defined as:

(*i*) Any loss that interrupts the continuity of production.

(*ii*) A loss of asset availability.

(*iii*) The non-availability of equipment.

(*iv*) A deviation from the *status quo*.

(*v*) Lack of meeting target expectations.

(*vi*) Any secondary defect etc.

8.12 TYPES AND CAUSES OF FAILURE

Basic types of failure are discussed below:

1. **Fatigue failures:** When fatigue loads are applied, such as in an automotive connecting rod, the fatigue strength of the material comes into play. When the fatigue strength of the material is exceeded, a crack can develop. This fatigue crack slowly propagates until a fracture occurs. The face of a fatigue failure tells us about the types and the magnitude of the load.
2. **Corrosion:** Corrosion significantly reduces the fatigue strength and increase probability of failure. Corrosion is result of the chemical attack on parts/materials by hostile fluids or atmospheres. Symptoms of corrosion are reddish brown area on rolling elements, raceways, cages and other components. As a result of corrosion, vibrations increases followed by wear with subsequent increase in clearance.
3. **Wear:** A variety of mechanisms results in the wear and tear or loss of material which may lead to failures.
4. **Contamination:** A grave premature bearing failure may be caused due to contamination. Dents or scratches in the bearing raceways and balls (or rollers) are symptoms of contamination which may result in vibration and wear. Airborne dust, dirt or abrasive substances are glaring examples of contaminants.
5. **Stress concentration:** It is a physical or metallurgical condition that increases the local stress in the component by some factor, like a small or no radius in a machined component, where the dimension changes. It can immensely effect in crack initiation which may consequently produce local stress.

6. **Hydrogen embrittlement:** If tensile stresses applied to hydrogen embrittled component may fail it prematurely. These hydrogen embrittlement failures though unexpected may sometimes become catastrophic. During material processing, hydrogen may get introduced into the material. Hydrogen embrittlement may occur without an external load or at loads, even when it is below yield stresses.
7. **Ductile overload failure and brittle overload failure:** When there is a great deal of distortion of failed part, it known as ductile failure. A ductile part may fail when it distorts and can no longer carry the needed load like an overloaded steel coat hanger which may break it into two pieces.

   A brittle fracture occurs when an overloaded part breaks with no apparent distortion. It may happen, if the material is very brittle, for example grey cast iron or when a heavy shock or extremely rapid load is applied to a normal ductile part. In a brittle overload failure, separation of the two halves or more pieces is not quite instantaneous.

8.13 FAILURE ANALYSIS (FA)

It is essential to identify and analyze the probability, causes and consequences of failures. It involves a logical and systematic investigation of various reasons of failures.

In a big industry, visual inspection, electrical testing, non-destructive evaluation, destructive evaluation may be adopted for failure analysis. There are various modes of failure analysis, for example, Proper Failure Mode Determination, Fracture Analysis, Fatigue Analysis, Overloads, Ductile and Brittle Failures, Chemical Attacks, Corrosion and Contamination Analysis, Stress Corrosion Cracking, Particle Analysis, Filter Residue Analysis, Welding, Soldering and Brazing Analysis, Vibration Analysis, Manufacturing Problem Analysis, Reverse Engineering of Components, Material Selection or Processing or Design Problem, Evaluation of Life Cycle Test Failure, Engineering Calculation, Atmospheric Testing, Operational Dynamic Analysis

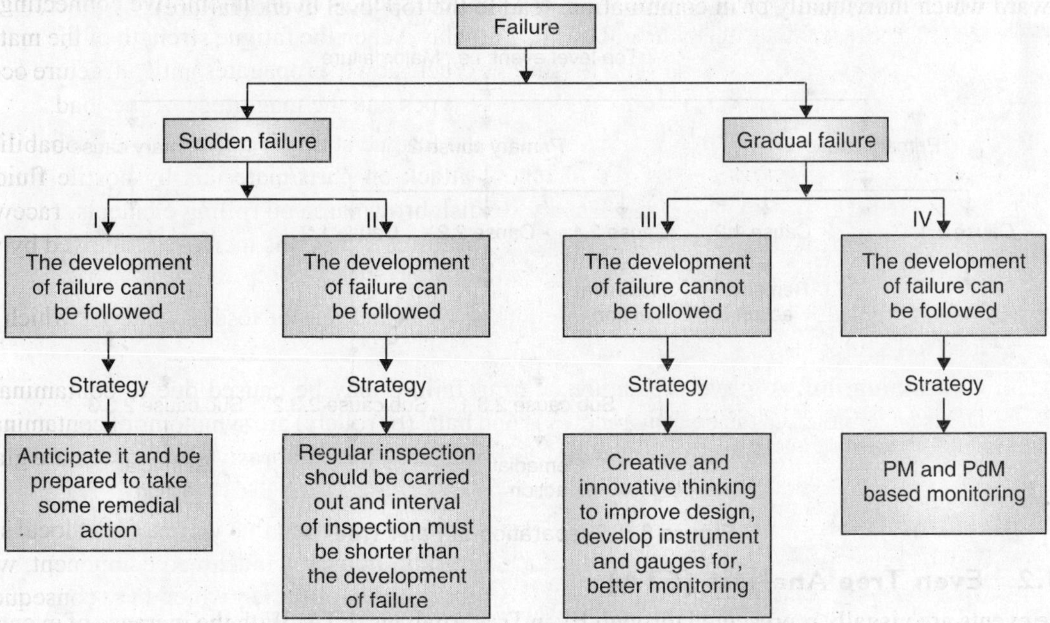

Figure 8.4 Failure Analysis vis-à-vis Maintenance Strategies

Figure 8.4 shows the failure analysis vis-à-vis various maintenance strategies. The objective of this failure analysis should be to bring the failures from left to right side, *i.e.*, from sudden to gradual and from "the development of failure cannot be followed" to "the development of failure can be followed". The ideal situation of maintenance in an industry may be achieved when most of failures fall in "IV" category and least in "I" category.

The following are the different failure analysis methods:

(*a*) Fault Tree Analysis (FTA)
(*b*) Even Tree Analysis (ETA)
(*c*) Route Cause Analysis (RCA)
(*d*) Root Cause Failure Analysis (RCFA)
(*e*) Failure Modes Effect Critically Analysis (FMECA)
(*f*) Finite Element Analysis (FEA)
(*g*) Electrical Stress Analysis
(*h*) Weibull Analysis or Life Data Analysis

8.13.1 Fault Tree Analysis (FTA)

Fault Tree Analysis (FTA) technique is adopted for reliability and safety analysis. By FTA, the state of the system (top event) in terms of the states of its components and sub-components (basic events) is established as shown in Figure 8.5.

The successive causes are analyzed, both quantitatively and qualitatively, and necessary remedial actions are planned to eliminate those causes. Boolean logic theory or Probability theory may also be used for analysis.

As a logical and structured process, it can help identify potential causes of system failure before the failure actually takes place. Explosion, toxic release, breakdown and malfunction are grave faults which are labelled as the top level event. By relating sequences of events, the fault tree is constructed downward which individually or in combination, lead to the top level event (failure).

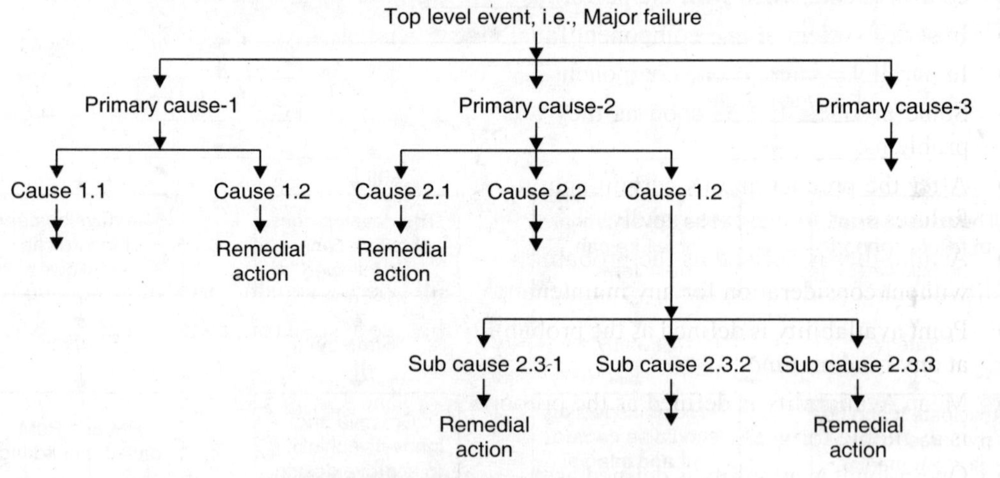

Figure 8.5 Preparation of Fault Tree

8.13.2 Even Tree Analysis (ETA)

All the events are visually represented through Even Tree Analysis, ETA. With the increase of event, the picture fans out like the branches of a tree. Event tress can be used to analyze both the systems like the one where all its components are continuously operating or some or all of the components are in standby mode. The starting point known as the initiating event disrupts normal system operation. The event tree displays the sequences of events involving success and/or failure of the system components.

8.13.3 Electrical Stress Analysis

The components that are likely to fail due to operating conditions are looked through Electrical Stress Analysis. It is based upon components application, input and operating voltages and currents, voltages and current peaks and surges, junction temperatures and temperature rise. Analysis results can be shown in a tabular form that will depict each component, values of actual operating parameters and safe limit for those parameters. This table highlights the overstressed components. This analysis is very common in maintenance practice.

8.13.4 Weibull Analysis or Life Data Analysis

The process of discoveries the trends in products or system failure data, by statistical means and using them to predict is failures in similar situations is termed as Weibull Analysis. By finding out these trends, we can attempt to correct or compensate for them, so as to improve product reliability. It may be applied in a different of fields, practices and disciplines, as it has been the most widely used distributions for failure data analysis. This type of analysis is especially used in for mechanical, electrical, chemical, electronic, material and human failure analysis.

SUMMARY

- Reliability Engineering deals with life data analysis of products like equipments, components, systems and processes etc.
- Reliability refers to the fact that a product function without failure for desired periods during their whole design life from conception to junking.
- Reliability is concerned with performance of a product during its entire lifetime while quality control is concerned with the performance of a product at manufacturing stages.
- In series system, if one component fails, the system will fail.
- In parallel system, if one component fails, the system does not fail.
- Some products fail as soon as they are put in service due poor design or manufacturing problems.
- After the product attains certain age, it enters the period where it begins to wear out, and failures start to increase rapidly.
- Availability is defined as the probability that a product when used under stated conditions, without consideration for any maintenance, shall operate satisfactorily at any given time.
- Point availability is defined as the probability that the system (or product) will be operational at any random time.
- Mean Availability is defined as the proportion of time during a time period when the system is available for use.
- Operational availability is defined as the measure of availability that includes all experienced sources of down time, logistic downtime, etc.
- Maintainability is the ease in time or resources with which a system or product can be modified to correct faults, bring performance to peak or adapt to a changed environment.

- Failure means lack of success, deterioration, decay or stoppage etc of an item for any reasons.
- When there is a great deal of distortion of failed part, it known as ductile failure.
- A brittle fracture occurs when an overloaded part breaks with no apparent distortion.
- Failure Analysis is essential to identify and analyze the probability, causes and consequences of failures.
- In fault tree analysis, the state of the system (top event) in terms of the states of its components and sub-components (basic events) is expressed.
- The components that are likely to fail due to operating conditions are looked through Electrical Stress Analysis.
- The process of discoveries the trends in products or system failure data, by statistical means and using them to predict is failures in similar situations is termed as Weibull Analysis.

ASSIGNMENT

1. Students should be given know how of some of the important softwares related to Reliability Engineering.

EXERCISES

1. Explain the term 'Reliability' and also explain the need of reliability.
2. Differentiate between reliability and quality.
3. Explain bath tub curve.
4. What are different types of reliability? Explain any two of them.
5. How reliability of a system can be improved?
6. What do you understand by reliability program?
7. What are advantages of reliability?
8. Explain the availability and its classification.
9. What do you understand by maintainability? What are its advantages and disadvantages?
10. What is failure in maintenance? What are basic reasons of failure in mechanical items?
11. What is failure analysis? Name the various techniques of failure analysis and explain briefly any two techniques?
12. Give some idea of Weibull Analysis in relation to maintenance.

❑❑❑

Maintenance

9 CHAPTER

9.1 NECESSITY OF TESTING, REPAIR AND MAINTENANCE (TRM)

The terms testing, repair and maintenance are inter-related to each other. These activities are performed separate department known by Maintenance department. The significance of testing, repair and maintenance is given below:

(*i*) When a new industry is set-up, we required to carry a thorough checking and testing of machines, equipments and other accessories to ensure that they have been supplied as per order (requirement) or not.

(*ii*) Testing, repair and maintenance helps to reduce scrap and minimize wastage. Thus, brings control in quality.

(*iii*) The testing is also required to check, if the machinery is emitting hazardous, fumes, gases or other pollutants. Thus, pollution control is obtained.

(*iv*) To ensure the safe working conditions of machinery and equipment, they must be tested from time-to-time to check the load conditions and their performances or failure.

(*v*) A maintenance engineer with his expertise in machinery, their testing, upkeep and maintenance can offer reliable advice to management in their decision on replacement or discarding of obsolete machinery, if required.

(*vi*) The plant, machinery, building and installations are assets of the company. They are a great trust on the management, which is fulfilled by regular testing, repairs and maintenance.

(*vii*) A progressive management always plans occasional programs for shop reorganization, expansion involving shifting and reinstallation of old machinery or even reconditioning of plant.

(*viii*) For the statutory norm equipments, which are dangerous to human life, there is a necessary requirement of testing at regular intervals or before the start of operation. For example, boiler cranes, safety valves, engine emissions etc.

9.2 INTRODUCTION TO MAINTENANCE

The discipline of maintenance engineering refers to an application of concepts of engineering to the optimization of equipments, procedures and departmental budgets to procure better maintainability, reliability and availability of equipment.

Maintenance includes a combination of a number of preventive or corrective activities carried out to retain a product or item in more consumer friendly service period thereby restoring it to acceptable conditions. Its study is becoming more viable functionally because of rapid rise in equipments, systems, machines and infrastructures.

With the onset of Industrial revolution, devices, equipments, machines and structures have become increasingly complex, requiring a large number of personnel, vocations and related systems to maintain them.

The chief objectives of maintenance are as follows:

(*i*) To keep the plant, items, machines and equipments fit and that too at the lowest possible cost.

(*ii*) To reduce the breakdown probability.

(*iii*) To ensure the safe working conditions of the machines, equipments or plant, so as to prevent accident.

(*iv*) To bring the wastages and scrap to minimum and to achieve maximum quality control.

(*v*) To optimize their full use, machines and equipments should be kept in a prefect condition without interruption.

9.3 WHY IS MAINTENANCE REQUIRED?

Requirement of maintenance is based upon the following reasons:

(*i*) Machines and equipments are properly maintained to improve their working condition at all times. This ensures optimum utilization of production facilities and resources.

(*ii*) Their accuracy also improves.

(*iii*) The manufactured products are more cost-effective.

To achieve it, the machines need to be periodically inspected for

(*i*) Inadequate lubrication.

(*ii*) Machine level change.

(*iii*) Overloading of the machine tools.

(*iv*) Wear and tear of different parts.

(*v*) Dust, heat and humidity.

(*vi*) Mechanical vibrations.

(*vii*) Corrosion of different parts.

9.4 ROLE OF MAINTENANCE ENGINEER

A maintenance engineer is supposed to look after the following:

(*i*) The capacity assurance objectives of the organization. Developing the plans to achieve these objectives.

(*ii*) Resources and skill management to execute the plans.

(*iii*) Enhancing effective maintenance management systems.

(*iv*) Monitoring the progress of the plans.

(*v*) Improvement in the expected financial and operational benefits.

(*vi*) Optimization of maintenance costs.

(*vii*) Providing suitable advice on the design of new installations.

(*viii*) Leading the organization in a way that ensures continuous improvement since its inception.

9.5 REQUIREMENTS FOR MAINTENANCE

To achieve high degree of maintenance, the following essential requirements may be kept in mind:

(*i*) Effective supervision of maintenance department.

(*ii*) Procuring well-trained operators.

(*iii*) A good lubrication schedule should be chalked out.

(*iv*) Keeping an adequate stock of spare parts ready.

(*v*) Pollution free surroundings with proper ventilation.

(*vi*) Coordination of maintenance department with production, planning and purchasing departments.

(*vii*) Proper maintenance record should be maintained.

(*viii*) Proper consultation with the machine manufacturers whenever needed.

9.6 MAINTENANCE APPROACH

Maintenance is a continuous process and necessity. The approach varies from company to company, according to its organization and the type of production for which its plant is erected. The purpose remains avoidance of disruption of production with maximum efficiency. Multiple aspects of maintenance should be carried out routinely according to the manufacturer's manual like oiling and greasing. In online production industries, the routine servicing is an essential principle. It is to be noted that an unexpected failure of a single machine can disrupt the production line.

On both the high capital cost and individual machines, critical to a company's total production program and economy, special precautions are required to ensure that the machine continues production.

Maintenance is often neglected in small to medium range machines, employed in jobbing shops. Many companies employ personnel for routine lubrication service. If no such service is provided, the operators and supervisors are engaged to carry out this program.

9.7 ECONOMIC ASPECT OF MAINTENANCE

Maintenance is concerned with organization's policy with respect to the maintenance function of its equipments. The policy decisions are more of economical than technical nature. The basic requirement for achieving maintenance includes selection of the correct maintenance strategy. Areas often neglected include defining availability and reliability requirements, measuring performance correctly and even identifying assets properly. Application of wrong technique can waste time, money and resources.

The savings which can be made by following proper maintenance policy are:

1. Avoiding losses of output due to machine breakdown.
2. Reduction in the cost of repair.
3. Capital invested can be recovered more quickly.

9.8 MANPOWER PLANNING

Manpower planning is defined as a process by which the management ensures that the right number and the right kind of persons are selected at the right place and right time and are doing the right things for the achieving objectives of the organization.

Proper utilization of resources can increase productivity level. It will also reduce wastage of time, money, efforts and energies. Manpower planning is a continuous process, since the workers of an organization are often subjected to changes. The different managerial functions such as planning, organizing, directing and controlling are based upon the manpower. Thus, manpower planning helps in the implementation of all these functions.

Manpower planning is a dual process. It not only analyses the current human resources but also makes forecast of manpower required in future. It offers following advantages:

1. Shortages and surpluses of workers can be easily identified by manpower planning. Hence, required action can be taken.
2. All the fresh recruitments and selection committees are based on manpower planning.
3. It also helps in reduction of the labor cost as excess staff can be identified and consequently overstaffing can be avoided.
4. It also helps in identifying available talents and accordingly training program can be conducted.

9.9 MATERIAL MANAGEMENT

Material management is the process by which an organization is supplied with the required material. Material management is defined as planning for the movement of materials and associated flow of information. It also covers the identification of material, procurement of material, cataloging of purchased material, standardization of material/product, storage and quality control of purchasing.

In most of the industries, material is purchased and transported to the plant, material is then changed into finished product and this product is supplied to the customer. All these activities are associated with materials management. These activities of materials management are performed by the material manager. The functions of material management include procurement of materials, its distribution, warehousing, material planning, storage and disposal. It also deals with economic aspect of purchasing and the requirements of customers. Generally, material manager faces challenge of maintaining a consistent flow of materials required for production. An effective material management must have good support of quality management.

9.10 TYPES OF MAINTENANCE

The following two methods are adopted for maintenance:

(*i*) Breakdown Maintenance

(*ii*) Preventive Maintenance.

In breakdown maintenance, the repair may be done due to the breakdown while in the case of preventive maintenance, it is done due to prediction on the basis of periodic checking.

9.10.1 Breakdown Maintenance

Breakdown of a machine may be caused by the following twin reasons:

(*i*) If there is unpredictable failure of components which is hard to prevent.

(*ii*) On account of gradual wear and tear of its parts. It may be avoided by regular inspection. With timely periodic checking any part may be replaced, so as to avoid breakdown.

Evidently in breakdown maintenance, parts may be replaced or defects may be removed only when the machine can no longer perform its function. The production department is compelled to get the repair done. After removal of the defect, the maintenance engineers attend the machine only when another failure occurs in future.

Repairs in this case are done only on failure, which disrupts the whole production, during an important work. Due to increase of depreciation cost, payment to idle operators is costly, if an emergency repair is called for.

9.10.2 Preventive Maintenance

It is interchangeably termed as planned maintenance, scheduled maintenance or systematic plant maintenance. This is an important function as it reduces maintenance cost, keeps the equipment in good operational condition and thus, increases its reliability.

Preventive maintenance locates the cause of trouble and removes them before the breakdown occurs. Thus, it justifies the idea that prevention is better than cure. Imperative safeguard against costly breakdown is to inspect, lubricate and check the equipment before time as frequently as possible. In order, to utilize the equipment to the full in reliable condition, necessary measures may be taken to maintain it and to prevent overloading, dampness, negligence and its misuse. Frequent inspection may be done on the basis of its usefulness, importance and delicacy. Thus, periodic inspections or assessment modes help to find out the reasons, leading to breakdown and to rectify them, when they are in initial stages. These repairs may be done whenever anticipated. This will have the least effect on the production schedule. In fact, this type of repair needs less time as compared to that of breakdown repair. The preventive maintenance actually saves time and increases production.

9.11 PREVENTIVE MAINTENANCE CARD

Preventive maintenance can be carried out weekly, monthly or annually. Card or Chart is used to maintain records of preventive maintenance check. There are three types of cards for preventive maintenance:

(*i*) **Weekly Preventive Maintenance Card:** It includes:

(*a*) Cleaning of machine/ equipment

(*b*) Check for lubricating oil

(*c*) Check for coolant

(*d*) Filter check

(*e*) Minor adjustments (like tightening of nuts/bolts etc.)

(*ii*) **Monthly Preventive Maintenance Card:** It includes the works done in weekly maintenance and following additional work:

(*a*) Belt change

(*b*) Check for all axis movement and alignment

(*iii*) **Annual Preventive Maintenance Card:** It includes the works done in monthly maintenance and following additional work:

(*a*) Machine level check

(*b*) Replacement of lubricating oil

The following steps are to be followed by the operator while performing preventive maintenance on the lathe

(*a*) Cleaning of machine

(*b*) Clean the chips from the machine surfaces

(*c*) Check the level of the cutting fluid reservoir

(*d*) Lubricate the bed ways daily with light oil

(*e*) Check the oil level in headstock periodically

(*f*) Check the apron oil level regularly

The various cards of Preventive maintenance for lathe machine are shown below:

<table>
<tr><th colspan="6">Weekly Maintenance Card</th></tr>
<tr><td colspan="6">Name of the Department:</td></tr>
<tr><td colspan="2">Name of the Machine:</td><td colspan="2">Model/Make:</td><td colspan="2">Year:</td></tr>
<tr><td colspan="6">For the .... week of .... month of .... year</td></tr>
<tr><td colspan="6">Name of the operator:</td></tr>
<tr><td>Job description</td><td>Lead Screw</td><td>Head Stock</td><td>Saddle</td><td>Speed box</td><td>Motor bearing</td></tr>
<tr><td>1. Lubrication
2. Check for coolant
3. Filter check
4. Minor adjustments</td><td></td><td></td><td></td><td></td><td></td></tr>
<tr><td colspan="6">Checked by ....</td></tr>
</table>

<table>
<tr><th colspan="3">Monthly Maintenance Card</th></tr>
<tr><td colspan="3">Name of the Department:</td></tr>
<tr><td>Name of the Machine:</td><td>Model/Make:</td><td>Year:</td></tr>
<tr><td colspan="3">For the .... week of .... month of .... year</td></tr>
<tr><td colspan="3">Name of the operator:</td></tr>
<tr><td>Job Description</td><td colspan="2">Operation Performed</td></tr>
<tr><td>1. Belt Change
2. Check for coolant
3. Check for movement of axis
4. Check for alignment of axis</td><td colspan="2"></td></tr>
<tr><td colspan="3">Checked by ....</td></tr>
</table>

Annual Maintenance Card		
Name of the Department:		
Name of the Machine:	Model/Make:	Year:
For the .... week of .... month of .... year		
Name of the operator:		
Job Description	Operation Performed	
1. Spindle axis alignment 2. Machine level check 3. Spindle bearing adjustment 4. Replacement of lubricating oil		
Checked by		

9.12 ORGANIZATION STRUCTURE OF MAINTENANCE DEPARTMENT

The buildings, plant and services of any organization structure are labelled as fixed assets and in many companies they form at least 50% of the money invested.

Table 9.1 Organization Structure of Maintenance Department

Maintenance Superintendent				
			Engineering Assistant	
Facilities Foreman	***Foreman Shop Maintenance***	***Foreman Planning and Scheduling***	***Foreman Engineering***	***Field Foreman***
1. Steam 2. Power 3. Water 4. Air	1. Maintenance 2. Repair 3. Lubrication	1. Work order System 2. Planning and Estimating 3. Scheduling 4. Backlog Control 5. Performance Report	1. Engineering 2. Design	1. Buildings 2. Yards 3. Fire 4. Protection 5. Disposal

In any company, big or small, some department of the main organization is made responsible for maintaining these important assets. The section or department busy in looking after the equipments, building etc., is called maintenance department. An organization structure is needed for the proper functioning of the maintenance department.

9.13 PROCEDURES ADOPTED IN MAINTENANCE

Maintenance is subject to multiple factors. Everyday maintenance may be done by the operators themselves. Before starting the work of their shift, cleaning, oiling and greasing is supposed to be done by the operators. The manufacturers issue maintenance instructions for his machines, which have to be strictly adhered to.

Preventive maintenance also depends largely on the operators. Therefore, as far as possible, one operator is assigned one machine. If the same machine is used for more than one shift, there has to be one operator for each machine for each shift. This system has many advantages.

The specially trained maintenance crew generally manages the periodic maintenance. The period decided on the basis of the past experience. The operating instructions have to be carefully followed by the operator. If any abnormal sound is heard or behaviour is noticed in the machine, the operator should immediately inform the maintenance crew about it to correct the defect immediately. It should not be neglected or postponed for a later date. Therefore, inspection and lubrication schedules should be strictly followed including the inspection of tools, accessories and other equipments of the machine.

9.14 CENTRALIZED AND DE-CENTRALIZED MAINTENANCE

In centralized maintenance, that persons having charge of maintenance of equipments, machines, building etc. have a common supervisor. Thus, the organization does not waste time and money on duplication of work. Here, workers are not involved in decision making.

Decentralized maintenance provides each and every area with its own maintenance force. The decentralized units may report to the maintenance supervisor, but they are individually responsible for maintenance in their own units. Decentralized maintenance allows the workers of the organization to become more familiar with the problems of a particular unit of the organization. Here, workers are involved in decision making.

9.15 RECENT DEVELOPMENTS IN MAINTENANCE ENGINEERING

With the rapid and radical growth in the field of diagnostic/predictive system, it is now easy to predict the failure of the machine in advance even before its occurrence so that maintenance staff may be prepared well to arrange necessary parts required and expert man power for upcoming repairs to expedite the operation of the machines.

1. **Mechanisms to monitor mechanical failure**

   (*i*) *Ultrasonic flow detector:* It is used to find out the appearance of minute cracks.

   (*ii*) *Strain gauge:* It is used to measure deflection of structural parts like columns and foundation.

   (*iii*) *Shock pulse meter:* It points out defects of rolling elements like ball bearing.

   (*iv*) Vibration meter: It assesses the vibration level to find out if it is within safe limit or not.

2. **Mechanisms to monitor thermal failure**

   (*i*) *Thermovision:* It scans the surface of a particular piece of equipment with the help of an infra red temperature sensor and afterwards displays on the video screen.

   (*ii*) *Infrared thermometer:* By the use of remote sensing, it detects hot spots in bearings or other parts of the machine.

3. **Mechanisms to monitor chemical failure**

   (*i*) *Corrosion meter:* The corrosion rate of chemical equipments can be measured by using this instrument.

   (*ii*) *Ultrasonic flow meter:* In order to determine the extent of choking in pipes or heat exchangers, this instrument is used.

(*iii*) *Ultrasonic leak detector:* It detects leak of gases at high pressure which cannot be otherwise seen or heard.

(*iv*) *Ultrasonic thickness meter:* It is used to measure thickness of paints, coatings etc. It also measures thickness of pipes and tubes in order to find out the wear rate.

4. **Development of more advanced techniques**

Following are some of the advanced techniques:

(*i*) Chromium plating for restoration of worn out shafts and bearings.

(*ii*) Plasma spraying of metals/alloys and other refractory materials. This process helps to restore worn out parts to original dimensions. It also serves to increase resistance in corrosion and erosion.

(*iii*) *Metalock system:* It is useful to repair cracks in heavy machines and prevention of its further spreading.

(*iv*) *Eutectic welding:* It is applied for deposition of hard surface materials on worn-out parts. They are later on machined to required dimensions and tolerances.

9.16 CONDITION MONITORING BY NDT

Condition Monitoring (CM) is the process of determining condition of machines or equipment, such as any change is indicative of development failure. It is an important component of predictive maintenance technique. The application of condition monitoring allows maintenance to be planned so that required action may be taken in order to avoid failure.

Machines with defects are more likely to fail than defect free machines. Once a defect has been identified, the failure development has already started and condition monitoring can only measure the deterioration of the condition. Intervention in the initial phases of deterioration is more cost effective than allowing the machine to fail completely. In today's competitive world, significant savings can be achieved by application of CM.

Non-destructive testing (NDT) includes wide group of analysis techniques used in industry to determine the properties of a material, or system without causing any destruction. NDT technique saves money as well as time. Also, it does not change the properties of the system under testing. Common NDT methods include ultrasonic, magnetic-particle, liquid radiographic, visual inspection eddy-current testing etc. It finds wide applications in field of engineering and medicine.

Based on NDT, some important types of condition monitoring techniques are:

(*i*) Infrared thermography

(*ii*) Alignment of shafts using LASER

(*iii*) Oil analysis

(*iv*) Leakage detection

(*v*) Ultrasonic crack detection

(*vi*) Ultrasonic thickness monitoring

Infrared Thermography: Infrared thermography is also called as thermal imaging. It detects infrared radiations emitted from object, converts it to temperature and displays image of temperature distribution. Temperature rise can be sources of number of problems in any machine or equipment. Condition monitoring by using this technique gives a quick and safe way of detecting problems in such situations. At the same time, it can be carried out while machine is under operation. Temperature measurement can be done from a distance without coming in contact with the machine.

Alignment of Shafts using LASER: Perfect alignment of the shafts is very important for high speed rotating machine since misalignment can cause failure of machine. LASER shaft alignment system utilizes LASER-based technology for checking alignment of shafts. LASER shaft alignment systems with wireless Bluetooth operation, color graphics and infinite battery life are available. Such system offers accurate and precision alignment of shafts within hundredths of mm.

Oil Analysis: Contamination in oil can be detected by means of condition monitoring by oil analysis. There can be different reasons of contamination such as presence of water particles, carbon deposits, traces of other oil etc. One of the techniques of condition monitoring through oil analysis is oil circuit and transformers where condition of the oil is monitored for the presence of carbon deposits.

Leakage Detection: Leakage is unintentional discharge of gases from mechanical equipments such as pressure vessels, pipelines, pressure containment parts, aerosol containers etc. Detection of colorless and odorless gases is a tedious task. The purpose of leak detection systems is to detect and locate leakage. Such system provides an alarm and displays other important information. Leaks can be detected by using liquid/gas penetrant techniques, electronic listening devices or bubble tests etc.

Ultrasonic Crack Detection: Detection of minute cracks is a difficult task specially if the cracks are near other inhomogeneities, pores, material inclusions or geometric features. The laws of physics that govern the propagation of sound waves through solids is used to detect minute cracks. High frequency sound waves are reflected from crack in predictable ways, producing distinctive sound patterns which is displayed and recorded. Ultrasonic crack detection technique finds application in manufacturing industries involving welds and structural metals.

Ultrasonic Thickness Monitoring: Ultrasonic thickness monitoring is quite an effective technique adopted for finding out wall thickness of corroded metals like steam pipes, water pipes, storage tanks, pressure vessels etc. It is also used for determination of thickness of metals, plastics, ceramics and glass etc. Measurements can be made from single side only. Ultrasonic thickness gauges work by measuring how long it takes for a sound pulse to travel through material under test. The ultrasound waves inside a material will bounce back from the opposite surface like an echo. Thus, thickness measurements can be made instantly and accurately. Ultrasonic thickness measurements save material, time and costs.

9.16.1 Advantages of condition monitoring

The advantages of condition monitoring are as follows:

1. Maintenance can be planned
2. It maximizes availability and reliability of machine
3. It minimizes unscheduled downtime
4. Cause of failure can be analyzed
5. It improves operating efficiency
6. It reduces maintenance costs (better planning)
7. It improves customer relations

9.16.2 Disadvantages of condition monitoring

The disadvantages of condition monitoring are as follows:

1. It requires careful choice of the correct technique
2. Monitoring machine cost is high
3. Skilled personnel needed
4. Strong management commitment needed.

9.17 COMPUTERIZATION OF MAINTENANCE

In today's techno-savvy world, quick and economical communication, job preparation, information collection and maintenance are accomplished far more efficiently by means of computer. Computerization of maintenance is required for maintaining database of information about different maintenance operations of an organization. This information can be used by maintenance workers to carry out their duties in a more effective way. It is used to store records such as information related to machines that require maintenance and store rooms containing the required tools/spare parts. Computerization of maintenance also helps in planning suitable maintenance program, tracking repair work, taking quick decisions, conducting cost analysis and verification of regulatory compliance. Such information can also be used when dealing with third parties, like in a liability case where the information can serve as a evidence that proper maintenance had been carried out.

9.17.1 Computerized Maintenance Management System (CMMS)

The term used for computerized maintenance system is Computerized Maintenance Management Systems (CMMS). However, Computerized Maintenance Management Information Systems (CMMIS) is more preferred since today's systems by design and by use are not used to manage maintenance but rather to inform about maintenance. CMMIS allows easier access to large amounts of information reducing time and cost.

CMMIS is used to perform preventive maintenance, asset management, work order and inventory control. It can also produce status reports and documents giving details of maintenance work. CMMIS take care of scheduling jobs, assigning duties, materials management, spare parts management and tracking information related to the cause of the problem. It also maintains data about machines like specifications, warranty and service contracts. It maintains different important documents required for the processing of safety requirements. Thus, perform various functions and offers several advantages

The above mentioned functions are formed by softwares available in market. These softwares can be either web-based or LAN Based. With the help of these softwares, maintenance work can be performed in effective and efficient manner. CMMS/ CMMIS software must have various characteristics such as it must be a sound, comprehensive, on-line and user-friendly. The CMMIS software offers the following benefits:

1. Improves speed and accuracy of the process
2. Improves efficiency of maintenance resources and hence, lowers unit cost
3. Helps in proper utilization of resources
4. Provides better performance and product quality
5. Improves service to the customers
6. Improves asset reliability, capacity assurance and equipment uptime.

SUMMARY

- The discipline of maintenance engineering refers to an application of concepts of engineering to the optimization of equipments, procedures and departmental budgets to procure better maintainability, reliability and availability of equipment.
- Maintenance includes a combination of a number of preventive or corrective activities carried out to retain a product or item in more consumer friendly service period thereby restoring it to acceptable conditions.

- The objective of maintenance is to keep the plant, equipment and machines fit at the lowest possible cost.
- Maintenance is a continuous process and necessity. The approach varies from company to company, according to its organization and the type of production for which its plant is erected.
- Manpower planning is defined as a process by which the management ensures that the right number and the right kind of person are selected at the right place and right time and are doing the right things for the achieving objectives of the organizational.
- Materials management is the process by which an organization is supplied with the required material.
- In breakdown maintenance, the repair may be done due to the breakdown.
- In preventive maintenance, repair is done due to prediction on the basis of periodic checking.
- Everyday maintenance is done by the operators themselves. Before starting the work of their shift, cleaning, oiling and greasing is supposed to be done by the operators.
- Ultrasonic flow detector is used to find out the appearance of minute cracks.
- Thermovision scans the surface of a particular piece of equipment with the help of an infra red temperature sensor and afterwards displays on the video screen.
- Ultrasonic leak detector detects leak of gases at high pressure which cannot be otherwise seen or heard.
- Centralized maintenance means that those person having charge of maintenance of equipments, machines, building etc have a common supervisor.
- Decentralized maintenance allows the workers of the organization to become more familiar with the problems of a particular unit of the organization.
- Condition Monitoring (CM) is the process of determining condition of machines or system, such that any change is indicative of a failure development.
- Non-destructive testing (NDT) includes wide group of analysis techniques used in industry to determine the properties of a material, or system without causing any destroying.
- Based on NDT, some important types of condition monitoring techniques are infrared thermography, alignment of shafts using LASER, oil analysis, leakage detection, ultrasonic crack detection and ultrasonic thickness monitoring
- Infrared thermography detects infrared radiations emitted from object, converts it to temperature and displays image of temperature distribution.
- Computerization of maintenance is required for maintaining database of information about different maintenance operations of an organization.

ASSIGNMENT

1. Discuss the importance of maintenance briefly.

EXERCISES

1. What is the testing of testing, repair and maintenance?
2. What is meant by maintenance? Is maintenance only a repair function?

3. Discuss the specific objectives of management.
4. What is the role of maintenance engineer is an industry?
5. Enumerate the chief requirements of good maintenance.
6. Write notes on maintenance approach.
7. Discuss economics aspect of maintenance.
8. Write notes on manpower planning. What are its advantages?
9. What do you understand by material management?
10. Discuss the various types of maintenance.
11. What is meant by preventive maintenance check?
12. Draw weekly, monthly and annual preventive maintenance card for lathe machine.
13. Write an essay on the organization structure of maintenance department.
14. Write notes on maintenance procedure.
15. Differentiate between centralized and decentralized maintenance.
16. Describe the recent developments in maintenance engineering.
17. What is meant by conditioning monitoring by NDT?
18. Discuss various techniques of conditioning monitoring by NDT.
19. What are advantages and disadvantages of condition monitoring?
20. Discuss computerization of maintenance.

❑❑❑

Maintenance Planning, Storage of Parts and Replacement Analysis

10
CHAPTER

10.1 MAINTENANCE OF MACHINE PARTS

In the following headings, maintenance of various machine parts like belt drive, chain drive, gear drive, shaft coupling and threaded fasteners is discussed.

10.1.1 Maintenance of Belt Drive

Belt is one of the most common and effective device used for transmission of motion and power from one shaft to the other by means of a thin inextensible band (belt) running over pulleys. Figure 10.1 shows belt drive.

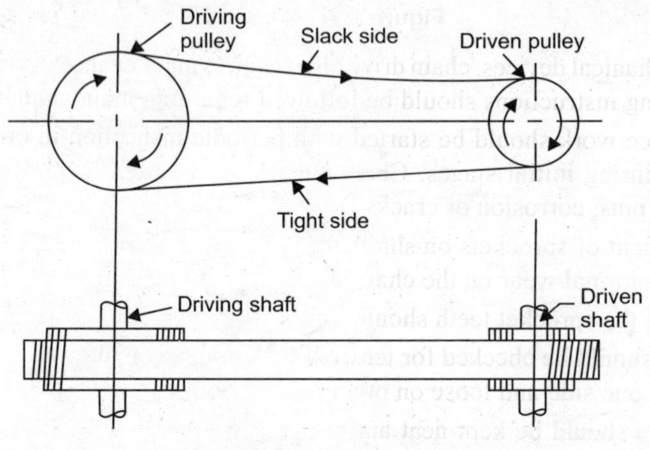

Figure 10.1 Belt Drive

A well-designed and maintained belt drive is capable of operating for many years when used under normal working conditions. The belt drives, which operate at high speeds, heavy loads, frequent stop/start conditions and at high working temperature require special attention. Some of the points related to maintenance of belt drive are as follows:

(*i*) Maintenance work should be started with periodic inspection. The alignment and tension of belt should be carefully checked.

(*ii*) Belt drive should be inspected for wear or rubbing against drive components.

(*iii*) Belt should also be checked for damage. It should be free from oil, grease, debris and dust. Any accumulation on belts can damage it.

(*iv*) Drive components such as bearings, shafts, motor also requires frequent inspection.

(*v*) The operating speed of belt drive should also be checked regularly. The working temperature also affects the life of belt. Higher working temperature reduces life of the belt.

(*vi*) Belt drive should be inspected for abnormal sound and mechanical vibrations. A properly maintained belt drive will operate smoothly and quietly without any vibrations.

(*vii*) Belt drive guard should not be loose or damaged.

(*viii*) The working environment in and around the belt drive should be safe. The working area should be free from debris and obstructions. Floors should be neat and clean.

10.1.2 Maintenance of Chain Drive

Chain drive is shown in Figure 10.2. It is used for transmission of power from shaft to the other by means of chain and sprockets. It is used in cycles, motorcycles and many machines.

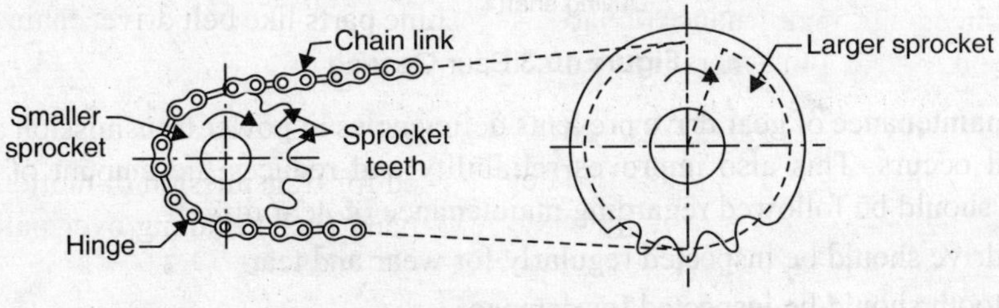

Figure 10.2 Chain Drive

Like other mechanical devices, chain drive also requires maintenance in order to obtain satisfactory working. The following instructions should be followed regarding maintenance of chain drive:

(*i*) Maintenance work should be started with periodic inspection in order to identify potential problems during initial stages. Chain should be checked for physical damage like broken part, loose pins, corrosion or cracks.

(*ii*) The alignment of sprockets on shafts should be proper. Any misalignment of sprockets can result in abnormal wear on the chain link plates and on the sides of the sprocket teeth.

(*iii*) The side of the sprocket teeth should be regularly inspected for signs of wear.

(*iv*) The chain should be checked for tension. It should never run with both sides tight. It should be tight on one side and loose on other side.

(*v*) Chain drive should be kept neat and clean. Any foreign particle on the chain can result in abrasive wear. It should be lubricated as and when required.

(*vi*) The shafts should be parallel to each other. They should be supported by strong bearings in order to avoid any displacement during operation.

(*vii*) The sprockets should be installed on the shaft according to the manufacturer's guideline.

(*viii*) Chain should be inspected for flexibility and elongation.

10.1.3 Maintenance of Gear Drive

Gears are basically toothed wheels, which are used to transmit power and motion from one shaft to another by means of successive engagement of teeth. They can be used to drive shafts that are parallel, intersecting or neither parallel non-intersecting. Preference to gear drive is given over belt drives and chain drives, where moderate or large amount of power has to be transmitted at a constant velocity ratio. Different types of gears available are: spur gear, helical gear, bevel gear and worm gear. Spur gearing is shown in Figure 10.3.

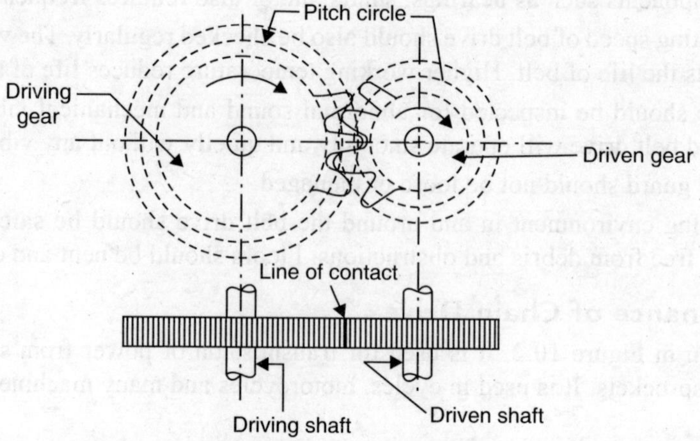

Figure 10.3 Spur Gearing

Regular maintenance of gear drive prevents deficiencies in power transmission and prevents the damage before it occurs. This also improves reliability and reduces the amount of downtime. The following points should be followed regarding maintenance of gear drive:

(*i*) Gear drive should be inspected regularly for wear and tear.

(*ii*) Gear tooth should be inspected for damage.

(*iii*) Gear should be checked for physical damage like crack or corrosion.

(*iv*) Gear must be inspected for surface damage like pitting. Gear must be grinded and polished properly in order to avoid pitting, which usually occurs in initial stages of operation.

(*v*) They should be properly and regularly lubricated with proper gear oil. This will protect them from corrosion.

(*vi*) The lubricant should be used as per manufacturer's recommendations.

(*vii*) The gear should be regularly cleaned using various washing liquids. Regular cleaning of the gear removes contaminants and maintains their performance. Cleaning also enhances life of the gear.

(*viii*) Gear drive should be checked for abnormal sound and excessive mechanical vibrations.

10.1.4 Maintenance of Coupling

Couplings are used to connect two shafts together at their ends for the purpose of transmitting power from one shaft to another. It also provides for shaft misalignment and introduces mechanical flexibility. The various couplings are:

(*i*) Rigid Coupling
  (*a*) Sleeve Coupling
  (*b*) Flange Coupling

(*ii*) Flexible Coupling
  (*a*) Oldham Coupling
  (*b*) Universal Coupling
  (*c*) Elastomeric Coupling

Figure 10.4 shows flange coupling. The various reasons for coupling failure are improper installation of coupling, poor coupling selection, operating beyond its capabilities.

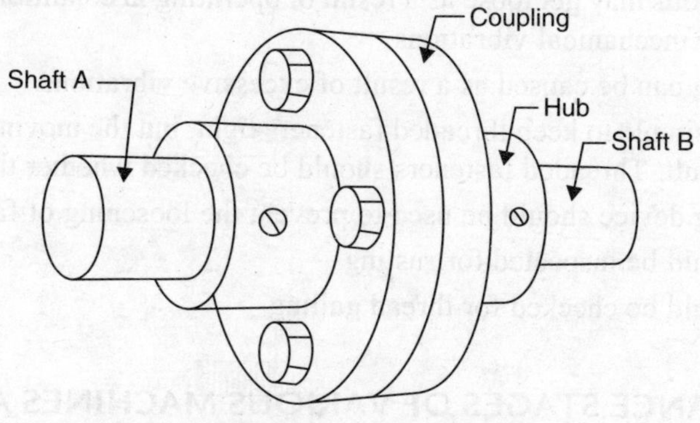

Figure 10.4 Flange Coupling

Regular maintenance of coupling should be performed in order to obtain satisfactory performance. Some of the important points related to maintenance of coupling are:

(*i*) Coupling should be inspected for abnormal sound and mechanical vibrations.

(*ii*) Inspection should also be carried out to check excessive heat generation and lubricant leakage.

(*iii*) Infrared thermography should be performed to measure the heat of the coupling. Heat may be generated as a result of misalignment, axial movement or torsion.

(*iv*) Operating speed must be checked. Proper operating speed should be maintained for longer life of coupling.

(*v*) It is necessary that gaskets be placed in the female coupling so tight connection is achieved. Gaskets should be inspected regularly for the signs of wear or cuts.

(*vi*) Visually inspection of the coupling must be carried out in order to check wear, corrosion and cracking. Coupling should be cleaned regularly.

(*vii*) Lubricant should be used as per manufacturer's recommendations.

(*viii*) Precision shaft alignment must be carried out.

10.1.5 Maintenance of Threaded Fasteners

Threaded fasteners are pieces of hardware having internal or external threads. They are employed to assemble different parts and facilitate easy disassembly. The common types of threaded fasteners are screw, nut, stud and bolt. An external thread is a thread on the outside of a shaft. The threads on a bolt are external threads. An internal thread is a thread on the inside of a hole. The threads in a nut are internal threads. Nut and bolt are shown in Figure 10.5.

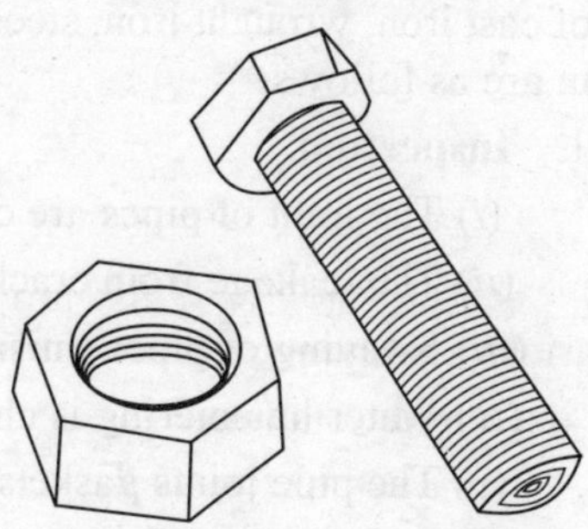

Figure 10.5 Nut and bolt

The following points should be remembered regarding maintenance of threaded fasteners:

(*i*) There are three modes of threaded fastener failure. Failures through the shank or threaded section of the fastener, external thread damage and internal thread damage. Fasteners should be properly installed as improper installation can damage threads. Use of incorrect fastener size can also damage thread.

(*ii*) Threaded fastener should be checked for wear as the wear of fastener material can take place as a result of repeated use.

(*iii*) The nuts and bolts may get loose as a result of operating in conditions such as varying stress, temperature or mechanical vibrations.

(*iv*) Thread fretting can be caused as a result of excessive vibrations.

(*v*) It seems very simple to keep threaded fasteners tight, but the moving nature of the machines makes it difficult. Threaded fasteners should be checked whether they are tight or not.

(*vi*) Proper locking device should be used to prevent the loosening of fasteners.

(*vii*) Fasteners should be inspected for rusting.

(*viii*) Fasteners should be checked for thread galling.

10.2 MAINTENANCE STAGES OF VARIOUS MACHINES AND EQUIPMENTS

The various stages of cumulative maintenance jobs are known as maintenance stages. These stages are Inspection, Servicing, Repair and Overhauling (ISRO) and are described below:

(*i*) **Inspection (I):** Inspection of machine or equipment is carried out to know and sense the possible defects and check the performance of various machine parts.

(*ii*) **Servicing (S):** Servicing or tune up is done on machine to bring it back to its peak performance.

(*iii*) **Repair (R):** This refers to the time duration between a series of maintenance stages bringing it to the initial pattern of maintenance stages.

(*iv*) **Overhaul (O):** Overhaul stands for:

(*a*) A compressive disassembly of a machine.

(*b*) Cleaning, inspection and measurement of every part.

(*c*) Repair and replacement of worn out parts.

(*d*) Checking and repairing of the foundation of machine.

(*e*) Reassembly of machine.

10.2.1 Maintenance Stages of Pipes and Pipe Joints

As a necessary accessory, pipes are used to carry the fluid from one place to another. They can be made up of cast iron, wrought iron, steel, brass or copper. The stages which are performed while maintaining them are as follows:

1. **Inspection**

(*i*) The joint of pipes are checked.

(*ii*) The leakage from cracks and joints is to be checked.

(*iii*) Swaying of pipes (misalignment) is inspected.

(*iv*) Water hammering is checked.

(*v*) The pipe joints gaskets are checked.

(*vi*) Regular checking of pipes is essential to find out corrosion.

(*vii*) Tightness of bolts at joints is thoroughly checked.

2. **Servicing**

(*i*) Pipe joints are disassembled.

(*ii*) Pipe joints are cleaned and washed.

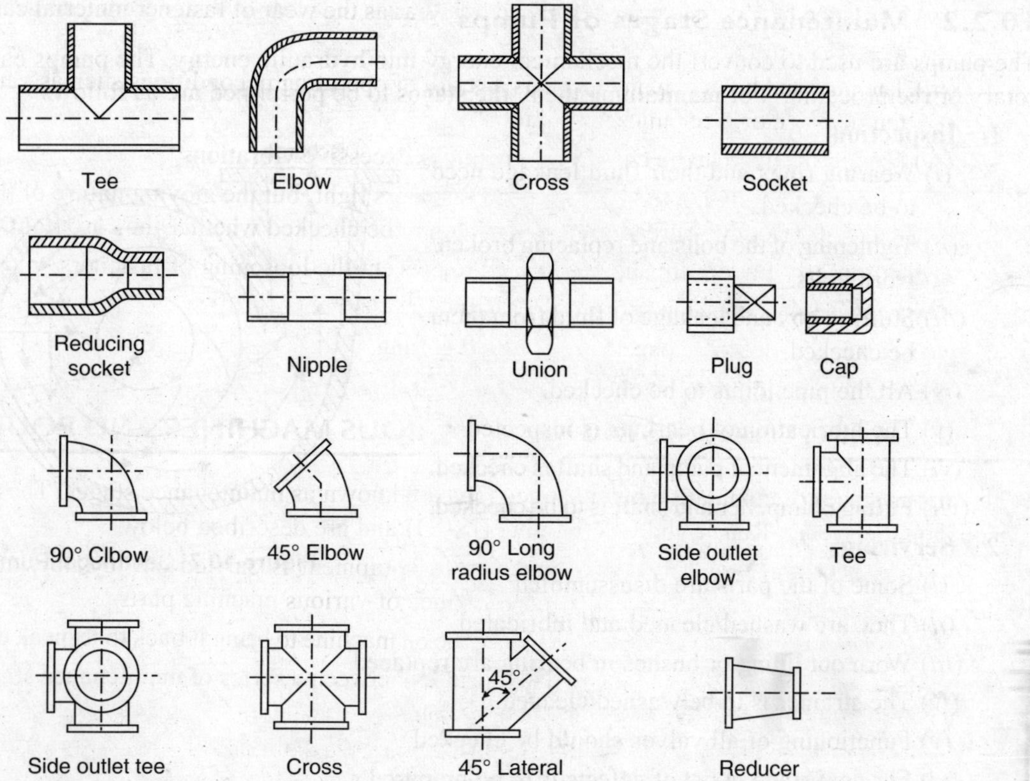

Figure 10.6 Various types of pipe joints

(*iii*) Alignment of pipes is checked and corrected, if required.

(*iv*) Gaskets are checked and replaced, if necessary.

(*v*) Due to temperature the expansion and contraction of pipe joints may take place, which should be checked.

3. **Repair**

(*i*) Pipes are checked for leakage and cracked pipes are replaced.

(*ii*) Worn out gasket be changed after tightening the bolts.

(*iii*) Iron cement is applied on the cracked portion of pipe.

(*iv*) Araldite may be applied to repair cracked part. A layer of glass may be used as top cover over it.

(*v*) All the parts of pipes are assembled.

4. **Overhaul**

(*i*) The pipe joints are disassembled.

(*ii*) All the joints (screwed) are lubricated.

(*iii*) Cracked pipes should be replaced by new ones.

(*iv*) Proper welding of the pipes may be done to form the pipe joint.

(*v*) Coating of glass fibre may be done to protect it from water or gas.

(*vi*) All the parts are assembled and accuracy should be checked.

10.2.2 Maintenance Stages of Pumps

The pumps are used to convert the mechanical energy into hydraulic energy. The pumps can be either rotary or reciprocating. For maintaining them, the stages to be performed are as follows:

1. **Inspection**
   (*i*) Wearing rings and their fluid leakage need to be checked.
   (*ii*) Tightening of the bolts and replacing broken/ worn bolts.
   (*iii*) Stuffing box and leakage of fluid from them be checked.
   (*iv*) All the pipe joints to be checked.
   (*v*) The lubrication of bearings is inspected.
   (*vi*) The alignment of pump and shaft is checked.
   (*vii*) Fitting of impeller and shaft is to be checked.

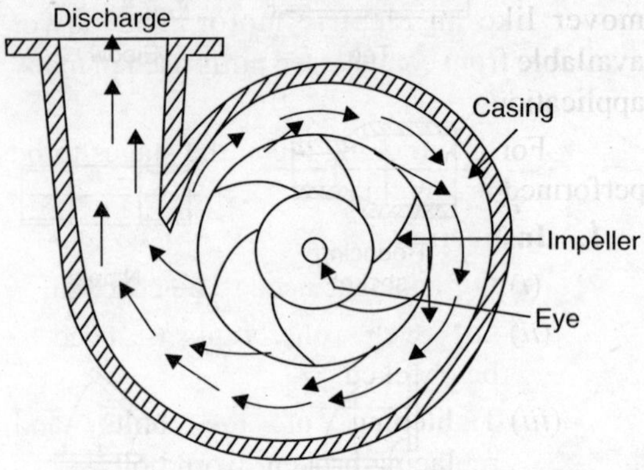

Figure 10.7 Centrifugal Pump

2. **Servicing**
   (*i*) Some of the parts are disassembled.
   (*ii*) They are washed/cleaned and lubricated.
   (*iii*) Worn out lining or bushes in bearing are replaced.
   (*iv*) The strainer is to be washed/cleaned.
   (*v*) Functioning of all valves should be checked.
   (*vi*) For next stage, a list of defects is to be prepared.
3. **Repair**
   (*i*) Major units and sub units linkages be disassembled.
   (*ii*) Worn out bearings should be replaced.
   (*iii*) The casing of the pump should be repaired, if required.
   (*iv*) All the nuts and bolts should be tightened. Worn out nuts and bolts should be replaced.
   (*v*) The coupling bushes, if necessary should be repaired/replaced.
   (*vi*) The relative position of parts in sub-assemblies be checked and fix them according to the drawings to give excellent packing.
   (*vii*) The worn out packing of stuffing box should be replaced.
4. **Overhaul**
   (*i*) The pump should be disassembled.
   (*ii*) All the parts should be washed and cleaned.
   (*iii*) The bearings and packing of pumps should be changed.
   (*iv*) Alignment of shafts is checked.
   (*v*) The worn out pipe fitting is replaced.
   (*vi*) After checking and repairing the foundation, the parts are painted and reassembled.
   (*vii*) The pump is run and its accuracy is checked.

10.2.3 Maintenance Stage of Air Compressors

An air compressor is a machine which compresses the air and increases its pressure. It inducts air from the atmosphere, compresses it and then delivers it to a storage vessel. From where, it is conveyed through the pipeline to a place where the supply of compressed air is required. The compression of air requires work to be done on it. Thus, compressor is always equipped with a prime mover like an electric motor. The power available from compressed air is used in many applications.

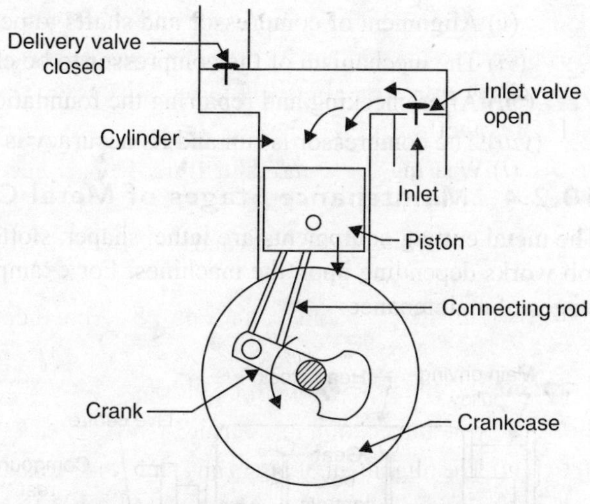

Figure 10.8 Reciprocating Air Compressor

For maintaining them, the stages to be performed are as follows:

1. **Inspection**
   (*i*) Oil leakages need to be checked.
   (*ii*) Oil level in the crankcase need to be checked.
   (*iii*) Tightening of the bolts and replacing broken/ worn bolts.
   (*iv*) All joints to be carefully checked for leakage.
   (*v*) Lubricating oil pressure must be checked.
   (*vi*) The alignment of compressor and shaft is checked.
2. **Servicing**
   (*i*) Some of the parts are disassembled.
   (*ii*) These parts are washed/ cleaned and lubricated.
   (*iii*) Surface of crank pin and journal should be checked.
   (*iv*) Clearance between the crank pin and main bearing should be checked and compared with prescribed data.
   (*v*) Cylinder, piston and piston rod need to be checked.
   (*vi*) Functioning of valves should be checked.
   (*vii*) For next stage, a list of defects is to be prepared.
3. **Repair**
   (*i*) Major units and sub-units should be dissembled. They are washed and cleaned.
   (*ii*) Worn out bearings should be replaced.
   (*iii*) All moving parts should be checked for wear and replaced, if required.
   (*iv*) All the nuts and bolts should be tightened. Worn out nuts and bolts should be replaced.
   (*v*) Functioning of valves should be checked.
   (*vi*) Lubricating oil should be drained out and fresh oil should be re-filled.
   (*vii*) The relative position of parts in sub-assemblies be checked and then fixed.
   (*viii*) Paint sub-units, if required before re-assembly.
4. **Overhaul**
   (*i*) The compressor should be disassembled. All the parts should be washed and cleaned.
   (*ii*) All the nuts and bolts should be tightened. Worn out nuts and bolts should be replaced.
   (*iii*) Worn out parts to be replaced.
   (*iv*) Lubricating oil should be drained out and fresh oil should be re-filled.

(*v*) Alignment of compressor and shafts to be checked.

(*vi*) The mechanism of the compressor to be checked for its relative motion.

(*vii*) After checking and repairing the foundation, the parts are painted and reassembled.

(*viii*) The compressor is run and its accuracy is checked.

10.2.4 Maintenance Stages of Metal Cutting Equipment

The metal cutting equipments are lathe, shaper, slotter, drilling machine etc. The stages have different job works depending upon the machines. For example, the lathe machine goes through the following stages of maintenance:

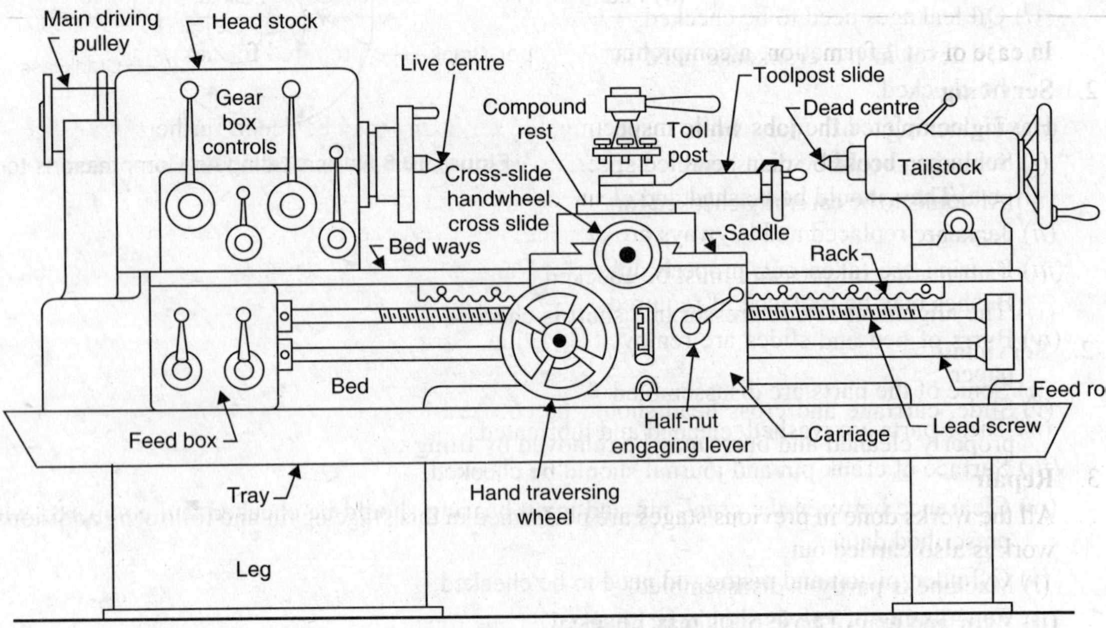

Figure 10.9 Lathe Machine

1. **Inspection**

(*i*) Thorough inspections of abnormalities in operation are carried out. Wear and tear is also checked.

(*ii*) Nuts and bolts are tightened and replaced, if damaged.

S.No.	*Parts*	*Maintenance Jobs*
1.	Beds and Slides	(*i*) The guide ways should be cleaned. (*ii*) Thin lubrication film is put on the guide ways. (*iii*) Coolant and burrs are also removed from it.
2.	Check/ Tool post	(*i*) Nuts are tightened. (*ii*) It is checked for security. (*iii*) It is properly cleaned. (*iv*) New packing may be provided, if tool chatters
3.	Head stock/Tail stock	(*i*) Surface is cleaned. (*ii*) Tapered holes are cleaned before the centers or tools are attached.

4.	Belt and Pulley	(*i*) Tension and alignment are checked. (*ii*) The burr fallen should be cleaned. (*iii*) Spilt lubrication oil should be cleaned.
5.	Springs	Tension should be checked.
6.	Gears	Noise and engagement should be checked.
7.	Lead screw	(*i*) The screw should be cleaned. (*ii*) The play in half nut is checked.
8.	Electric motor and starter	(*i*) Earthing, security cover and the pulley on its shaft are checked. (*ii*) Foundation bolts are checked and tightened, if required.

In case of fault formation, a comprehensive report should be prepared for next stage.

2. **Servicing**

Having completed the jobs while inspecting, the following may be added further:

(*i*) Subunits should be disassembled specially which need less technical knowhow such as tool post. They should be washed and cleaned.

(*ii*) Keys are replaced and keyways are cleaned.

(*iii*) Bearing are taken out, properly lubricated and adjustments are made for their alignment. Bushes may be replaced, if required.

(*iv*) Burrs of bed and slides are removed by filing. Scratches should be smoothened by emery paper.

(*v*) Slide, carriage and cross head should be checked for smooth movement. Ways should be properly cleaned and burrs may be removed by filing.

3. **Repair**

All the works done in previous stages are performed in this stage again and following additional work is also carried out:

(*i*) Machine is partly is disassembled.

(*ii*) Wear and tear of base parts may be checked and repair work may be done.

(*iii*) Worn out bushes should be replaced.

(*iv*) Wear on friction surfaces is to be measured and new linings is employed. Friction couplings should be replaced.

(*v*) The worn threads of lead screw may be machined and the burrs struck to it may be removed.

(*vi*) Worn out gears should be replaced and the gear oil should be changed.

(*vii*) The guard covers are replaced or repaired.

(*viii*) Worn our centre should be repaired or changed.

4. **Overhaul**

Operations done in previous stage are carried out once again and then all sub-units are dissembled. Each part is cleaned, washed, repaired and replaced, if necessary. The machine may be removed from the foundation and taken to the workshop for complete overhaul. Foundation is also checked and tested. Afterwards additional operations are done as follows:

(*i*) Before disassembly, precision and finish level are checked.

(*ii*) With the help of special tools, fixture and material handling devices, all the parts are to be dissembled.

(*iii*) All the parts are cleaned, washed and lubricated.

(*iv*) All the worn out parts are repaired or replaced.

(*v*) The guarding surfaces are cleaned, repaired or replaced. Finally, all the surfaces are painted.

(*vi*) The subunits are assembled and their relative motion is checked.

(*vii*) Finally, the complete machine is assembled.

(*viii*) The machine is put on its foundation and acceptance tests are carried out. Vibration analysis and surface finish of the job are evaluated.

10.3 MAINTENANCE OF MATERIAL HANDLING EQUIPMENTS

Material handling equipments are those mechanical equipments that are used for movement of materials in the process of manufacturing, distribution or disposal. The movement can be vertical or horizontal. The main functions of material handling equipments are as follows:

(*i*) To transport material from one point to another.

(*ii*) To position material at a single point so that it is in the correct position for subsequent handling, machining, transportation or storage.

(*iii*) To store materials over a period of time.

(*iv*) To collect and communicate the information which is used to coordinate the flow of materials within organization and between an organization and to customers.

Some of the material handling equipments are hand operated crane, forklift, hydraulic jack, mobile crane and winch etc. Maintenance of the material handling equipment is very important to keep them in good condition. Maintenance of different material handling equipments is discussed in the following headings.

10.3.1 Maintenance of Hand Operated Crane

Hand operated crane is shown in Figure 10.10. Regular inspections should be carried out to avoid crane failures, increases safety and reduces interruptions. By performing timely maintenance, unanticipated failures and emergency repairs can be avoided.

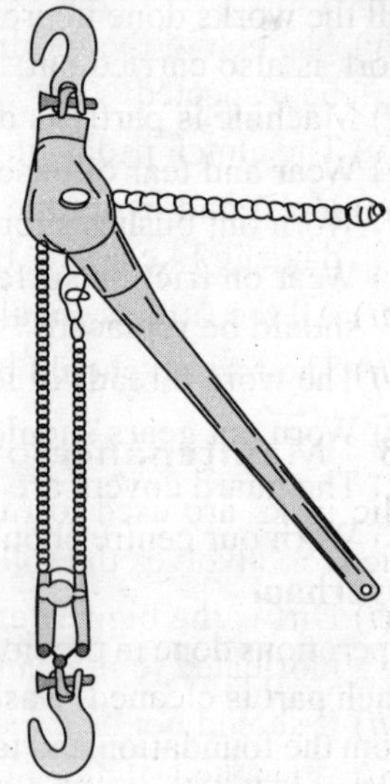

Figure 10.10 Hand Operated Crane

Following points should be kept in mind while carrying maintenance of hand operated crane:

(*i*) Company's requirements and manufacturer's manual should be strictly followed.

(*ii*) Maintenance work should only be carried out by competent person, selected through a formally documented assessment.

(*iii*) In order to find out whether the crane is performing the intended work or not, test should be carried out to check operation of each motion of the crane without load and with rated load applied.

(*iv*) Maintenance work should be carried out at a place where it can be carried out without any disturbance.

(*v*) Use maintenance checklists and enter all problems (inspections and repairs) in the logbook.

(*vi*) The structure of crane should be visually examined for physical damage such as crack, deformation and corrosion.

(*vii*) Check for loose nuts and bolts. Check for wear of brake and clutch system.

(*viii*) Working order of all mechanical components like gear box, hydraulic system etc should be checked.

10.3.2 Maintenance of Forklift

Forklift is shown in Figure 10.11. It is a basically truck with additional power than a normal truck. The forklift has forks to lift and transport materials. Proper procedures should be followed while performing maintenance work on forklift.

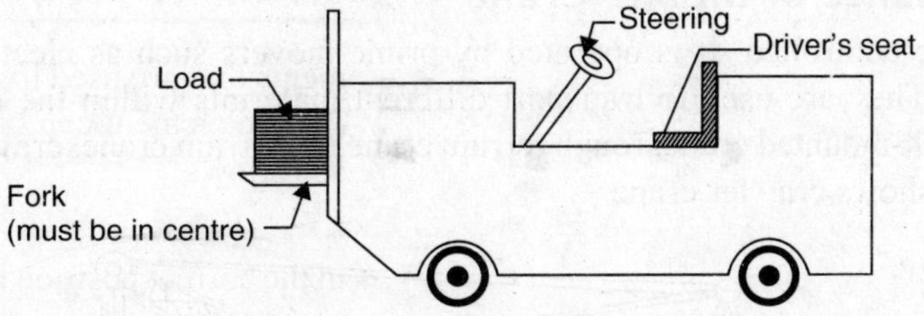

Figure 10.11 Forklift

Maintenance of forklift involves the following steps:

(*i*) Horn, safety warning devices, cab lights, outside lights and overhead guard should be checked for proper operation.

(*ii*) Wiring and battery compartment should be checked. Battery terminal and connectors should be cleaned. Level of distilled water in the battery should be checked. Grease should be applied on the battery connectors to prevent corrosion.

(*iii*) Lubricating oil should be changed regularly as per guidelines. Oil filter, fuel filter and the air filter should also be cleaned and changed, if required.

(*iv*) The belt of the forklift engine should be checked for cracks and frays. The drive axle should be inspected.

(*v*) The clutch pedal, steering cylinder and the drag links should be checked.

(*vi*) Hoist and hydraulic lines should be inspected. The forks should be inspected for physical damage like crack, bowing and bending.

(*vii*) All the fittings should be lubricated.

(*viii*) The forklift should be wiped down and spray cleaner should be used to the forklift.

10.3.3 Maintenance of Hydraulic Jack

Hydraulic jacks are used to raise heavy items. It uses incompressible hydraulic oil. Maintenance of hydraulic jack involves the following steps:

(*i*) Dirt is the biggest cause of failure of hydraulic system. Jack should be kept clean and well lubricated to prevent foreign particle from entering the hydraulic system.

(*ii*) It should not be exposed to rain, snow, sand or dust.

(*iii*) When hydraulic jack is not in use, piston and pump rods should be kept fully retracted. Jack should be stored on its base and in area free from corrosive vapors.

(*iv*) Jack should be checked for loose hardware and leakage of hydraulic oil.

(*v*) Hydraulic jack should be filled with approved hydraulic oils only. Manufacturer's recommendation should be consulted when changing the hydraulic oil. Brake fluid should never be used. Since brake fluid contains alcohol that will damage the internal seals.

(*vi*) Jack should be checked for physical damage like cracks.

(*vii*) Hydraulic jack should be checked for excessive wear, bending or other damage. Piston rod should be checked for damage.

(*viii*) Significant damage to hydraulic jack can occur by lifting a body too heavy for its rated capacity. Exceeding the lift capacity can place significant pressure on internal seals and valves. It will not only damage jack but also result in injury to the operator. So, rated capacity of jack should be kept in mind while lifting.

10.3.4 Maintenance of Mobile Crane

Mobile crane is cable-controlled. It is operated by prime movers such as electric motors or internal combustion engines. They are used to transport different materials within the industry. Examples of mobile cranes are truck-mounted crane, rough terrain crane, all terrain crane, crawler crane and floating cranes. Figure 10.12 shows crawler crane.

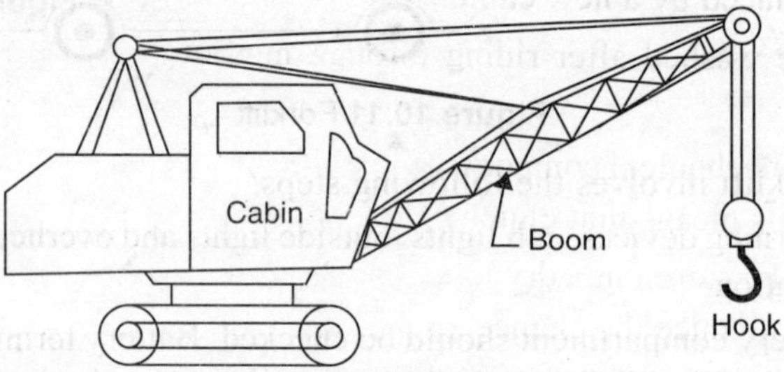

Figure 10.12 Crawler Crane

Following points should be kept in mind while carrying maintenance of mobile crane:

(*i*) The main objective of a maintenance work is to make sure that a mobile crane is safe to be taken into, or to continue in, use. Necessary inspection is carried out as a part of the maintenance work.

(*ii*) Company's requirements and manufacturer's manual should be followed. Use maintenance checklists and enter all problems (inspections and repairs) in the logbook.

(*iii*) Check for loose nuts and bolts. Tighten them, if required.

(*iv*) Different components should be checked for wear. Also, working of all mechanical components like gear box should be checked.

(*v*) Maintenance work on mobile crane should only be carried out by competent person, selected through a formally documented assessment.

(*vi*) Mobile cranes operate in quite risky environment which includes lifting heavy loads over people. It should be taken into account by the maintenance crew when determining the scope and nature of the maintenance. Maintenance work should be carried out at a place where it can be carried out without any disturbance.

(*vii*) In order to find out whether the crane is performing the intended work or not, test should be carried out to check operation of each motion of the crane without load and with rated load applied.

(*viii*) Crane should be examined for physical damage like crack, deformation and corrosion.

10.3.5 Maintenance of Winch

A winch is a mechanical device, driven by electric motor and used to pull heavy loads. Figure 10.13 shows electric winch.

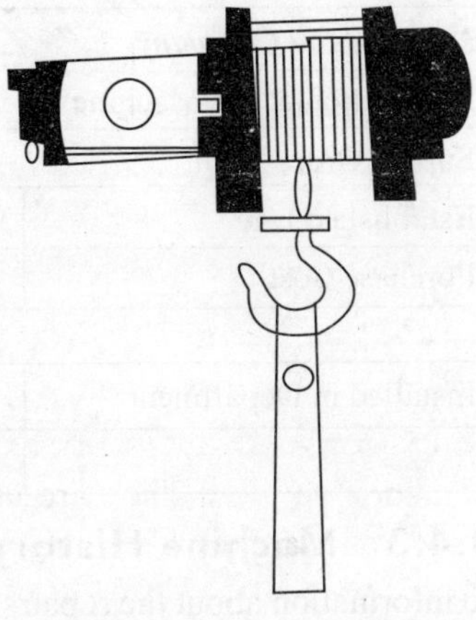

Figure 10.13 Electric Winch

Maintenance work should be carried out by competent person as per manufacturer's guidelines. Maintenance of winch involves the following steps:

(*i*) Winches should be inspected before using. Damaged winches and inoperable winches should be discarded immediately to avoid chances of accident.

(*ii*) Cable should be inspected for frays, cuts and abrasions. Inspection should be carried out before and after each use. If the cable is beyond repair, it should be immediately replaced by a new cable.

(*iii*) Cable should be washed after riding through mud or saltwater.

(*iv*) The working of mechanical component should be inspected. If the motor is running but drum is not turning, this means that clutch is not working properly.

(*v*) Winches should be stored in a dry location when not in use. Dirt in winch should be cleaned before storage to reduce the chance of corrosion.

(*vi*) Winches should be cleaned and lubricated. Dirt, dust, mud etc. can restrict its free movement.

(*vii*) The solenoid or remote control switch should be checked. If the motor is running in only one direction, the solenoid is defective or the remote control switch is inoperative.

Winches should never be overloaded. As overloading the cable can lead to failure.

10.4 MAINTENANCE RECORDS

Maintenance is explained in terms of the collection of data and its analysis. Maintenance department should maintain a log book to record data. This department maintains the following documents:

1. Instruction, Manuals and Drawings
2. Machine Data Sheet
3. Machine History Card.

10.4.1 Instruction Manuals and Drawings

The manufacturer provides them with the machines. They contain necessary and important information regarding the installation, operation, maintenance and testing of the machine. The maintenance engineer employed the firm refers to this booklet, if fault appears in the functioning of the machine.

10.4.2 Machine Data Sheet

Full record of every machine is maintained by the maintenance department in form of machine data sheet. It contains the information as depicted in Table 10.1.

Table 10.1 Machine Data Sheet

Name of the Company		*Card No.*
Description of the Machine		Name of the Department
Supply Order No.		Address of the Supplier
Established Life	Date of Receipt	Date of Installation
Purchase Cost	Cost of Installation	Total Cost
	Details of Transfer	
Installed in Department	Transferred from section with date	Transferred to section with date
	Specifications of the Machine	

10.4.3 Machine History Card

The information about the repairs, modifications, replacements and overhauling conducted on the machine is maintained in this card. It means that card contains the record of happenings in the life of the machine. It helps the maintenance department to find out detailed behaviour of the machine to further analyze the life expectancy of the component and to decide modification. The data available on the history card is used to find out the causes of failures of machine.

Table 10.2 Machine History Sheet

Name of the Company		Card No.			
Name and Description of the Machine		Reference No. Instruction Manual No. Equipment Record No. Drawing No.			
Identification No.		Location			
Date	Details of maintenance carried	Down Time	Repair Time	Parts Replaced	Date of Next Replacement

10.5 STORAGE OF PARTS FREQUENTLY USED FOR REPLACEMENT

Following points should be followed for the storage of parts frequently used for replacement:

(*i*) The parts which are frequently used should be stored in a space which is easily accessible.

(*ii*) Such parts are usually handled again and again. They should be handled with care in order to prevent from physical damage.

(*iii*) Proper handling equipment should be employed.

(*iv*) Storage system should be designed properly.

(*v*) Parts packed in large cartons should be stored on pallets in a walk-in pallet rack on the ground or stored directly on the ground. Cartons can be picked by a forklift or by a worker directly.

(*vi*) The place of storage should have clear marking. Parts should be stored in orderly manner so that they can be easily located when required.

(*vii*) The place of storage should be free from obstructions. Movement of worker, material and handling equipment should be easily accessible.

(*viii*) Storage place should be free from waste, grease, nails, metals chips or oil. Waste material should be removed regularly. Proper housekeeping is a must.

10.6 STORAGE OF PARTS NOT EASILY AVAILABLE IN MARKET

Following points should be followed for the storage of parts that are not available in the local market:

(*i*) Special attention must be paid to the parts which are not easily available in market. As missing part can result to a standstill of entire industry and cause significant financial loss.

(*ii*) Some parts can be stored for a number of years and defective parts can be returned. Optimized logistic processes make sure that the replacement parts reach their destination as quickly as possible.

(*iii*) Efforts should be done to recondition or re-use the part instead of discarding it as scrap. This will help to a great extent in that case when parts are not easily available in the market.

(*iv*) In some cases, the parts which are not easily available in the market are manufactured in that industry itself. This ends depends on the market.

(*v*) In some cases, cannibalization of parts can be done which refer to reuse of salvageable parts.

10.7 INTRODUCTION TO 5-S PRACTICE

As a tool to ameliorate the work-environment and ambience, the Japanese are well known for their "5-S" practice. Due to its operational usefulness and utility, many countries have started using this technique. It is not only useful for improving the work environment, but also for improving total quality management processes. The "5-S" practice can be used for solving day-to-day problems. The words "5-S" includes the following five Japanese words:

(*i*) Seiri

(*ii*) Seiton

(*iii*) Seiso

(*iv*) Seiketsu

(*v*) Shitsuke

Their English equivalents are Structurise, Systemise, Sanitise, Standardize and Self-Discipline. We can say that these "5-S" terms mean Organization, Systematic, Cleaning, Standardization and Discipline.

Seiri (Organization): It is the first practice of "5-S" practice. Organization means sorting items (like tools, components and materials), which are required for a particular job from those that are not. Efforts should be made to select the few important items and place them at convenient place while discard wanted items. Waste basket should be place within 4 steps of the worker for throwing away thrash.

Seiton (Systematic): It is the second practice of "5-S" practice. It means that all the items should place in a systematic order. It reduces the time wasted in locating and placing things in an organization. It emphasizes on clear designation of names and places

Seiso (Cleaning): It is the third practice of "5-S" practice. In an organization, cleaning means allocation of responsibilities. Every worker of the organization should be completely aware of his duties and responsibilities. By doing so, all responsibilities are absolutely clear and there will be no undefined or unallocated area.

Seiketsu (Standardization): It is the fourth practice of "5-S" practice. This practice emphasizes on strict and continual maintenance of organization. It also includes standardizing the principles of "5-S" practice. By maintaining standardized conditions, worker can be motivated to act quickly in any condition

Shitsuke (Discipline): It is the fifth and last practice of "5-S" practice. It is concerned with maintaining and following "5-S" principle. It involves development of the ability of doing things in accordance to the rules and regulations. This will helps in developing habit of framing and following rules.

10.8 FIFO (FIRST IN FIRST OUT) PRACTICE

Some material deteriorates with passage of time. Therefore, it is necessary not to overstock such materials and adopt FIFO (First In First Out) practice. This practice, material first placed in store is used before more recently acquired materials. In order to achieve this, the following systems are employed:

(*i*) **Double area system:** In this system, double area is set for each item. The new lots are placed in the other empty area, and required materials are used from old lot.

(*ii*) **Moving Division system:** In this system, area set for each item is somewhat larger than required to provide space between new and old lots. When material becomes old, it is moved to right side from where it is taken out and new is placed on the left side.

(*iii*) **Gravity feed system:** In this system, new lot is placed on the top of the piles whereas old is removed from the bottom. For example coal, which is handled in this way.

10.9 REPLACEMENT ANALYSIS

Financial running of an industry to great extent depends upon its replacement decision. An industry makes three types of major replacement decisions like:

(*i*) Replacement of a machine, when it wears out or becomes obsolete.

(*ii*) Another machine is needed for expansion.

(*iii*) New machine substitutes the old one.

Machines are great tools of production. An industry wants to have maximum profit on its product in a highly competitive market. Therefore, a decision to replace a machine is to be done carefully after considering many factors.

10.10 REASONS FOR REPLACEMENT OF EQUIPMENTS

There can be many causes to replace equipments/machines like:

(*i*) **Working conditions:** Replacement decision is taken on the basis of the functioning capacity of equipment or machine. Equipments or machines which create noise, are causing pollution, or are accident prone should be replaced.

(*ii*) **Obsolescence:** New machine or equipment comes in the market, which is capable of efficiently producing more products of good quality using less labour, it becomes imperative to replace the existing machine or equipment, although it had no operational or functional defect.

(*iii*) **Inadequacy:** With the change of product design to meet the customers demand or quantity to be manufactured, the equipment or machine becomes inadequate and therefore, it should be replaced.

(*iv*) **Deterioration:** Due to wear and tear, if a machine or equipment does not function properly, it certainly needs to be replaced. Faulty equipments produce low quality products. It decreases efficiency and profit and increases its cost and maintenance.

10.11 GUIDELINES IN REPLACEMENT ANALYSIS

Following principle may be used as guidelines for replacement analysis:

1. **Equipment in operation**

   (*a*) *Consider:*

   (*i*) Cost of operation

   (*ii*) Maintenance and repair cost

   (*iii*) Down time cost

   (*iv*) Salvage value

   (*b*) *Need not consider:*

   (*i*) Original cost

   (*ii*) Money spent on its repair and maintenance

   (*iii*) Unrealistic book value

2. **New equipment**

   (*a*) *Consider:*

   (*i*) Initial cost

   (*ii*) Interest on capital investment

   (*iii*) Salvage value

   (*iv*) Cost effectiveness of improved product

   (*v*) Savings on labour charges

   (*b*) *Need not consider:*

   (*i*) Savings not obviously assessable

   (*ii*) Overhead charges.

10.12 METHODS FOR REPLACEMENT STUDIES

Following methods are usually employed for equipment replacement studies:

(*i*) Pay Back Period Method.

(*ii*) Total Life Average Method.

(*iii*) Annual Cost Method.

(*iv*) Present Worth Method.

(*v*) Rate of Return Method.

(*vi*) Machinery Allied Product Institute method.

10.12.1 Pay Back Period Method

This method shows how long it will take (in years) to pay back invested capital.

Let,

P: Pay back period (in years)

C: Original capital investment

R: Annual return expected (*i.e.*, total annual earning after deducting taxes)

Then,

$$P = C/R$$

This method is not so reliable as it lacks dependence on insurance interest and maintenance.

10.12.2 Total Life Average Method

In this method, the grand total of the costs spent on buying, operating and maintaining equipment is obtained and which is divided by the total estimated life to get an average annual cost.

10.12.3 Annual Cost Method

In this method, a comparison of the annual costs of obtaining service from different equipments/ machines is done as follows.

Annual cost = Capital recovery + Operating costs

where, capital recovery (C.R.) is,

$$\text{C.R.} = (P - \text{L}) \times \text{Capital recovery factor} + L \times i$$

$$\text{C.R.} = (P - L) \times \text{C.R.F.} + L \times i$$

$$\text{Capital recovery factor (C.R.F.)} = \left[\frac{i \times (1+i)^n}{(1+i)^n - 1}\right]$$

where,

P: Cost of equipment; *n*: Life of equipment in years

i: Rate of interest; *L*: Salvage value.

10.12.4 Present Worth Method

To properly evaluate the present value of new equipment/machine, this method gives most accurate and reasonable results. For the sake of comparison, future costs are translated into today's value of money in the following manner.

A "Present worth of rupee" is today's value of money of tomorrow's money (at certain interest rate). This can be further explained as follows:

If we have ₹ 5000 to invest and want to know what will be its worth in 10 years at 10% interest, using the formula

$$F = P\,(1 + i)^n$$

where,

F: Worth of money in future

P: Present amount of money

i: Rate of interest

n: Number of years

Substituting in equation, we get,

$$F = 5000\,(1 + 0.10)^{10} = 5000\,(2.594) = ₹\,12{,}970.$$

Thus, ₹ 5000 today is worth ₹ 12, 970 after 10 years from now or in other words, ₹ 12,970 after 10 years from now has a present worth of ₹5000.

If we compare the two alternatives, all the costs must be translated into present worth and they must be compared for equal length of services.

In above equation, $(1 + i)^n$ is called compound amount factor. The values of $(1 + i)^n$ can be seen from tables for equal payment series. Present worth factor (P.W.F) can be calculated by putting the values of n and i in equation.

10.13 SOLVED EXAMPLES

Example 1. *If the cost details of two machines be as follows, find out the most economical one.*

S.No.	*Factor*	*Machine I*	*Machine II*
1.	*Initial cost (P)*	*₹ 18,000*	*₹ 28,000*
2.	*Annual operating cost*	*₹800*	*₹500*
3.	*Life (n)*	*6years*	*6 years*
4.	*Interest rate (i)*	*5%*	*5%*
5.	*Salvage value (L)*	*₹1000*	*₹4000*

Take, the value of CRF for i = 5% and n = 6 years as 0.1970.

Solution:

For Machine-I,

$$\text{Capital Recovery (C.R.)} = (P - L) \times \text{C.R.F} + L \times i$$

$$= (18000 - 1000) \times 0.1970 + 1000 \times \frac{5}{100}$$

$$= 17000 \times 0.1970 + 50$$

$$= 3349 + 50$$

$$= ₹\,3399$$

$$\text{Total Annual cost} = \text{C.R.} + \text{Annual Operating Cost}$$

$$= ₹\,3399 + 800$$

$$= ₹\,4199$$

For Machine-II,

$$\text{Capital Recovery (C.R.)} = (P - L) \times \text{C.R.F} + L \times i$$

$$= (28000 - 4000) \times 0.1970 + 4000 \times \frac{5}{100}$$

$$= 24000 \times 0.1970 + 200$$

$$= 4728 + 200$$

$$= ₹\,4928$$

$$\text{Total Annual cost} = \text{C.R.} + \text{Annual Operating Cost}$$

$$= ₹\,4928 + 500$$

$$= ₹\,5428$$

The total annual cost of Machine-I is less, here, Machine-I should be selected.

Example 2. *Initial cost of machine is ₹12,000, annual operating cost is ₹800 and life of machine is 8 years. If the salvage value is 1200 and interest of rate is 5%, find out the present worth of the machine.*

PWF_1 for n = 8 and i = 5% is 0.6768 (for salvage cost)

PWF_2 for n = 8 and i = 5% is 6.4632 (for annual operating cost)

Solution:

Initial cost = ₹ 12000

Present worth of the annual operating costs $= 800 \times PWF_2$

$= 800 \times 6.4632$

= ₹ 5170.56

Present worth of the salvage value $= 1200 \times PWF_1$

$= 1200 \times 0.6768$

= ₹ 812.16

Present worth of machine = ₹ (12000 + 5170.56 – 812.16)

= ₹ 16,358.4

SUMMARY

- Inspection of machine or equipment is carried out to know and sense the possible defects and check the performance of various machine parts.
- Servicing is done on machine to bring it back to its peak performance.
- Repair refers to the time duration between a series of maintenance stages bringing it to the initial pattern of maintenance stages.
- Overhaul stands for:
  (*a*) A compressive disassembly of a machine.
  (*b*) Cleaning, inspection and measurement of every part.
  (*c*) Repair and replacement of worn out parts.
  (*d*) Checking and repairing of the foundation of machine
  (*e*) Reassembly of machine.
- Maintenance is explained in terms of the collection of data and its analysis.
- Maintenance department maintains the following documents:
  (*a*) Instruction, Manuals and Drawings
  (*b*) Machine Data Sheet
  (*c*) Machine History Card
- The manufacturer provides instruction manual with the machines which contain necessary and important information regarding the installation, operation, maintenance and testing of the machine.
- Full record of every machine is maintained by the maintenance department in form of machine data sheet.
- Machine history card contains the information about the repairs, modifications, replacements and overhauling conducted on the machine is maintained in this card.
- Special attention must be paid to the parts which are not easily available in market. As missing part can result to a standstill of entire industry and cause significant financial loss.

- Some material deteriorates with passage of time. They should not be overstocked and FIFO (First In First Out) practice should be adopted for such material.
- Replacement decision is taken on the basis of the functioning capacity of equipment or machine. Equipments or machines which create noise, are causing pollution, or are accident prone should be replaced.
- With the change of product design to meet the customers demand or quantity to be manufactured, the equipment or machine becomes inadequate and therefore, it should be replaced.

ASSIGNMENT

1. Students should be asked to visit some local manufacturing industry to see various maintenance practices and tools used.

EXERCISES

1. Write notes on maintenance of belt drive.
2. Discuss maintenance of chain drive.
3. What points should be followed during maintenance of gear drive?
4. Describe maintenance of coupling.
5. Write note on maintenance of threaded fasteners.
6. Differentiate between inspection, servicing, repair and overhaul.
7. Specify the maintenance stages for pipes and pipe joints.
8. Explain the maintenance procedure for pumps
9. Discuss the maintenance stages of reciprocating air compressor.
10. Explain maintenance stages of lathe machine.
11. How maintenance of hand operated crane is done? Explain.
12. Describe maintenance of forklift.
13. What points should be followed during maintenance of hydraulic jack?
14. Describe the procedure of mobile crane maintenance.
15. How is maintenance of electric winch carried out? Describe.
16. Elucidate on maintenance records and its significance.
17. Explain the procedure of maintenance of equipment data sheet and equipment history card.
18. What points should be kept in mind for storage of parts frequently used for replacement?
19. Describe how the parts not easily available in market, stored?
20. Describe 5-S practice of storage.
21. What do you understand by FIFO practice?
22. Write about replacement analysis.
23. What are different reasons of equipment replacement? Discuss any two of them.
24. What are the guidelines in replacement analysis?
25. Discuss any two methods for replacement analysis.
26. Distinguish between pay back period and total life average method.

❑❑❑

Engineering Materials

11

CHAPTER

11.1 INTRODUCTION

The ever increasing fund of knowledge and information of materials and their properties is of great importance for engineering applications. Various products should be made up of such pertinent material which has properties suitable for the conditions it is to be used.

The engineering materials can be classified as:

(*a*) Metals and their alloys.

For examples iron, steel, copper, aluminium, magnesium, zinc, etc.

(*b*) Non-metals.

For example ceramics, polymers, composite, etc.

All the metals used in engineering works can be further classified into the following two categories:

1. **Ferrous metals:** The metals which have iron as their main constituent are called as ferrous metals although other constituents like carbon, sulphur, manganese, phosphorus, etc. may be present in varying proportions. For example cast iron, steel, etc. All ferrous metals are magnetic and give little resistance to corrosion.
2. **Non-ferrous metals:** The metals which have elements other than iron as their main constituent are called as non-ferrous metals like copper, aluminium, magnesium, zinc, etc. They are not magnetic and are usually more resistant to corrosion as compared to ferrous metals. Non-ferrous metals are expensive than ferrous metal, but their use in many cases is unavoidable due to special properties, which they possess like good heat and electrical conductivity, lightness in weight and resistance to corrosion.

The selection of material is one of the most difficult steps in the designing process of any product. The main objective of material selection is to minimize cost while meeting product performance goals. Availability of the appropriate and high quality material is another important issue. Systematic selection of the best material for a given application starts with analysis of properties and costs of potential materials. In order to select the proper material mechanical, electrical, thermal and chemical properties have to be considered.

11.2 MECHANICAL PROPERTIES OF ENGINEERING MATERIALS

Mechanical properties indicate material reaction to the application of mechanical forces and load. Following are some of the important mechanical properties.

1. **Strength:** It is the property of material by virtue of which it withstands load without fracture.
2. **Toughness:** It is the property of material, which enables it to withstand shocks and vibration.
3. **Stiffness:** It is the property of material by virtue of which it resists deformation under stress. It is also known as rigidity.
4. **Elasticity:** It is the property of a material by virtue of which it regains original shape after deformation when external forces are removed.
5. **Plasticity:** It is the property of a material by virtue of which it retains the deformation under load permanently.
6. **Hardness:** It is the property of material by virtue of which it resists abrasion and penetration.
7. **Ductility:** It is the property of material, which allows it to be drawn into thin wire by application of a tensile force.
8. **Brittleness:** It is the property of breaking of a material with little permanent distortion.
9. **Malleability:** It is the property of material, which allows it to be beaten into thin sheets.
10. **Creep:** It is a slow, time-dependent, progressive deformation under constant loading of the material.
11. **Resilience:** It is the ability of a material to absorb energy elastically. On removal of load, the energy stored is given off exactly as in a spring when the load is removed.
12. **Hot Hardness:** It is the ability of the material to remain hard even at high temperature.

11.3 OTHER IMPORTANT PROPERTIES OF ENGINEERING MATERIALS

Following are some of the important properties of engineering materials.

1. **Melting point:** It is the temperature at which state of material changes from solid to liquid. It is dependent on pressure and is specified at standard atmospheric pressure.
2. **Specific heat:** It is defined as the amount of heat per unit mass required to raise the temperature by unit degree.
3. **Thermal conductivity:** It is the ability of a material to conduct heat. Heat transfer across materials of high thermal conductivity will be higher and faster as compared to materials having low thermal conductivity.
4. **Thermal expansion:** When heat is added to materials, the amplitude of the atom's vibration also increases. This, in turn, increases the separation between the atoms causing the material to expand. If the material does not go through a phase change, the expansion can be easily related to the temperature change. The linear coefficient of thermal expansion describes the relative change in length of a material per degree temperature change.
5. **Electrical conductivity:** It is a measure of how well a material accommodates the movement of an electric charge.
6. **Thermoelectricity:** When two dissimilar metals are joined from end to end and their junction is maintained at different temperatures, a small voltage is produced and this is known as thermoelectric effect.

7. **Magnetic permeability:** It is defined as the ease with which a material can be magnetized. It is a constant of proportionality that exists between magnetic induction and magnetic field intensity.
8. **Magneto-striction:** It is the phenomenon of expansion or contraction of a material when it is subjected to magnetic field.
9. **Corrosive resistance:** It is defined as the resistance offered by a material to rusting and corrosion.
10. **Machinability:** It is defined as the ease with which a material can be machined.
11. **Weldability:** It is the ability of the material to be welded together by the application of heat and pressure.
12. **Castability:** It is the ease of forming a casting.

11.4 FERROUS METALS

The ferrous metals commonly used in engineering applications are pig iron, wrought iron, cast iron, carbon steel and alloy steel. The principal raw material for all ferrous metals is pig iron.

11.4.1 Pig Iron

The pig iron is the fundamental material from which all iron and steel products are derived. This raw material obtained from the chemical reduction (known as smelting) of iron ore in a blast furnace. The pig iron consists of carbon along with small portion of silicon, manganese, sulphur and phosphorous. The presence of carbon in pig iron is higher which makes it hard and brittle. Therefore, this iron is not suitable for making any useful item.

11.4.2 Wrought Iron

Wrought is tough, malleable, ductile and easily to weld. It has very low carbon content and has fibrous inclusions, known as slag. This slag of wrought iron is extremely useful in black smithy operations and imparts peculiar fibrous fracture to the material. The non-corrosive slag constituents provide it corrosive-resistance. Various antique examples of wrought iron can be noticed on balconies, doors and other architectural elements from as far back as the period of Romans. Presently, it is used to manufacture chains, hooks, railway couplings, nut and bolts, pipe fitting, etc.

11.4.3 Cast Iron

Basically cast iron is an alloy of iron and carbon. It is obtained by refining pig iron by re-melting it in a cupola furnace with limestone, used a flux which combines with some of the impurities of pig iron and removes it in the form of slag. Cast iron contains 2.5 to 5% carbon alongwith varying portions of silicon, sulphur, phosphorus and manganese. Different alloying elements like nickel, chromium, vanadium may also be present.

Cast iron is brittle. Hence, it cannot be used in those parts, which are subjected to shocks. The properties, which make it valuable, are its low cost, good casting characteristics, high compressive strength and excellent machinability. The quality of cast iron depends both on the percentage of carbon as well as the condition in which carbon is present. Carbon in cast iron usually exists in two forms:

(*a*) Free carbon or graphite

(*b*) Combined carbon or cementite.

The different types of cast iron in use are discussed below:

1. **Grey cast iron:** The grey color is due to the presence of free carbon (in form of graphite flakes). The graphite in its structure acts as a lubricant. The amount of carbon present is 3 to 3.5%. It has got high compressive strength, low tensile strength and no ductility. It has excellent machinability and good anti-friction properties. It is used in making machine bodies and cylinder blocks, etc.
2. **White cast iron:** As a evident from the name, white cast iron, has got a silvery white appearance. The carbon present is in the combined form and hence, is very hard and brittle. Due to hardness, it cannot be easily machined. It's resistance to wear is high. The amount of carbon present is 1.75 to 2.5%. It is used in making rim of car, railway brake blocks and rolls for crushing plates, etc.
3. **Mottled cast iron:** Mottled cast iron is the mixture of grey and white cast iron in composition, color and properties. The fractured surface is either white with grey specks or grey with white spots and patches. It is hard and brittle. It contains about 1.75% carbon. It is used for making lamp post, manhole covers and pipes, etc.
4. **Chilled cast iron:** It is a form of white cast iron produced by rapid cooling of molten iron. It has got high wear resistance. It is used for making wheels of railway carriage.
5. **Nodular cast iron:** In nodular cast iron, the graphite is precipitated in the form of nodules. This can be achieved by the addition of nickel-magnesium to molten grey cast iron. It has got high strength, toughness, wear resistance and machinability. It is used in making hydraulic cylinder, etc.
6. **Alloy cast iron:** Alloy cast iron contains alloying elements such as nickel, magnesium, molybdenum and chromium. The use of these alloying elements imparts special properties to it such as increased strength, higher wear resistance and improved corrosion resistance. It is used in automobile industries to manufacture cylinders, pistons, rings, etc.

11.4.4 Steel

Steel is an alloy of iron and carbon. The carbon content can be as high as 1.5%. The carbon is found in the form of iron carbide, because of the ability to increase the hardness and strength of steel. Silicon, sulphur, phosphorus and manganese are also present in small proportions. Steel combines good properties of strength in shear, tension, compression, malleability and ductility, hardness and toughness, thermal and electrical conductivity and machinability. There are two types of steel:

(*a*) Unalloyed steel or plain carbon steel or *simply* carbon steel

(*b*) Alloy steels.

11.4.5 Carbon Steel

It can be divided into three main groups according to percentage of carbon:

1. **Low/mild carbon steel:** It contains carbon from 0.05 to 0.3%. It has remarkably high elongation property. It is malleable and ductile. It can be forged and welded. It is used in the manufacture of wires, ropes, thin sheets, channels, plates for ships, bolts and rivets, etc.
2. **Medium carbon steel:** It contains carbon from 0.3 to 0.6%. It has high tensile strength. It is used for manufacturing reaction chamber for chemical engineering application, boiler drums, aero-engine cylinder, etc.
3. **High carbon steel:** It contains carbon from 0.6 to 1.5%. It is hard and less ductile. It cannot be forged or welded. It is used for making shear blades, chisels, punches and dies, etc.

11.4.6 Alloy Steel

Alloy steels may be defined as steels to which elements other than carbon are added in sufficient quantities to improve their properties. The important alloying elements are nickel, chromium, manganese, silicon, tungsten and vanadium, etc.

These elements may be either used separately or in combination to get required characteristics. Nickel improves toughness, strength and ductility. Chromium imparts corrosion resistance, strength, hardness and decreases machinablity. Manganese is added to improve strength and ductility. Silicon improves resistance to corrosion and imparts electrical properties. Tungsten increases ability to retain hardness at elevated temperatures. Vanadium helps in obtaining a fine grain structure. Cobalt gives hot hardness by retention of hard carbides at high temperature. Phosphorus makes the steel brittle and also improves resistance to corrosion.

1. **Stainless Steel:** Stainless steels contain chromium together with other elements such as nickel. It is a good resistant to the effects of acids, salts and alkalies. Stainless steel has good strength and toughness and has extremely good resistance to corrosion and heat. It is used for making hydraulic, steam and oil pumps, valves and household utensils, etc.
2. **Alloy Tool Steel:** Various alloying elements are added to plain carbon steel to impart desired properties to it. These steels have greater wear resistance and hot hardness than plain carbon steel. These are widely used for drills, taps, reamers, etc. Their hot hardness is sufficiently high, due to which they can be used in high speed turning or milling. Depending upon the method of quenching, these steels are of two types:

   Type O tool steel: These are hardened by quenching in oil. Its composition is 0.9% carbon, 1% manganese, 0.5% tungsten and 0.5% chromium. These steels are used in cold working applications like punching, blanking, shearing, forming and drawing dies, etc.

   Type A tool steel: These are hardened by cooling in the air. Type A tool steel contain higher amount of alloying elements as compared to type O steels. It contains 1% carbon and 5% chromium. These steels are used in cold working applications like thread rolling dies, coining dies and gauges, etc.
3. **High Speed Steel (HSS):** High speed steel (HSS) is basically high carbon steel to which various alloying elements like tungsten, molybdenum, chromium, vanadium and cobalt are added to improve hardness, toughness and wear resistance. These materials are deep hardening and can be quenched in oil, air or salt. HSS are capable of retaining their hardness upto 600°C and therefore they can be operated at higher cutting speeds as compared to alloy tool steels. Due to this, the name high speed steels is given to them. There are two types of high speed steels:

   Tungsten type steels (T-series): In this type, the tungsten is the main alloying element which is about 12 to 20%.

   Molybdenum type steels (M-Series): In this type, the tungsten is partially or completely replaced by molybdenum. It is cheaper in comparison to tungsten type and is more rapidly sharpened. It also has greater toughness at the same level of hardness.

   The popular 18-4-1 HSS contains 18% tungsten, 4% chromium and 1% vanadium. Cobalt may also be added.
4. **Free Cutting Steels:** It contains sulphur and phosphorus. These steels have higher sulphur contents than other carbon steels. The carbon content of free cutting steels varies from 0.1% to 0.45% and sulphur from 0.08% to 0.3%. These steels are used where rapid machining is the prime requirement.

11.5 HEAT TREATMENT OF STEEL

The importance of heat treatment is that, it changes the micro-structure of the materials and improves properties. Some of heat treatment processes for steels are discussed below.

Hardening

Hardening is the process to raise the temperature of steel above the critical temperature and cools it in a medium such as oil, water or brine. The temperature of hardening is dependent on the carbon content and it is about 760 to 800°C. The time duration of maintaining the temperature depends on the depth of hardening required. The degree of hardness can be varied by the rate of quenching. For low and medium carbon steel, water is used as a cooling medium. While for high carbon steel and alloys, oil is used. In order to avoid cracking, interrupted quenching should be followed.

Case Hardening

Case hardening is a diffusion process in which an alloying element (carbon or nitrogen) is diffused into the surface of metal. It is performed on low carbon steel to artificially enrich the surface with carbon upto a certain depth. Filing or drilling becomes very difficult after case hardening but it provides a hard surface with tough core. Also, the surface becomes resistant to wear after case hardening. Journals and gudgeon pins are case hardened.

Carburising

Carburising is a basically a procedure of case hardening. The object is placed in an iron box, packed with carbon rich environment, (charcoal) and is heated upto temperature of 900 to 950°C for certain time. The temperature is to be maintained for several hours depending on the thickness required. It is then followed by quenching, so that carbon is locked in the structure. It is possible to carburize only some portion of a part of the object. Several objects can also be packed together in a single box. The surfaces, which do not require carburising, should be kept covered. Mixture of asbestos powder and fire clay may be used for covering such surfaces or objects.

Cyaniding

Cyaniding is a superficial case hardening process. It is fast and efficient and is mainly used on low carbon steels. In this process, the object is immersed in a cyanide bath at about 850 to 950°C. It is then quenched and rinsed, in water or oil, to remove any residual cyanide. This process produces a thin layer on the surface of the object. Cyaniding improves hardness and wear resistance. It is used for small parts like bolts, nuts, screws, etc. The advantage of this process is that, if required, the bright finish of machined parts can be maintained during this process. The major disadvantage of cyaniding is that cyanide salts are poisonous.

Nitriding

Nitriding is the process in which, nitrogen is introduced to the steel. In this process, steel is maintained at 450-500°C in an atmosphere of ammonia gas and dissociated ammonia. The time duration in this environment determines the depth of penetration. The free nitrogen is formed during dissociation process. This nitrogen is very active and units with iron in the steel to form nitrate. The hardness is achieved by the formation of nitrides. The advantage of this process is that it causes little warping and distortion. The nitriding process is often followed by surface grinding to remove outer layer.

Flame Hardening

Flame hardening is a form of localized hardening. This process involves use of oxy-acetylene flame on the surface of the steel to be hardened. The surface is heated above the upper critical temperature followed by immediate quenching with jet of water. Gear teeth, crankshaft and cams are hardened using this method. This is basically surface hardening process which produces hard surface layer of about 2 mm to 6 mm depth while the internal parts retain original properties. The main difference between this process and other surface hardening processes is that the composition of the steel being hardened remains unchanged.

Induction Hardening

Induction hardening is based upon induction heating. It produces similar surface treatment regime as flame hardening. The steel component to be hardened is placed inside a water-cooled copper coil which has alternating current through it. This causes the outer surface of the component to heat up. The rate of heating and the depth of heating can be controlled by varying frequency and current. Thus, the depth of hardness is also controlled. This process is quite suitable for surface heat treatment. The advantage of induction hardening is that it takes few seconds. In a furnace, the same process can take several hours or even days.

Annealing

The internal stress developed during the operations such as welding, rolling or forging can be relieved by this annealing. This process involves heating the steel to a temperature in the range 550°C to 650°C and held at this temperature before being cooled slowly at a controlled rate. This reduces stresses by allowing dislocations to rearrange to a lower energy configuration. This process can be used to allow further forming operations and to prevent distortion of the steel components due to subsequent machining operations.

Normalizing

Normalizing is a process which is used to provide uniformity in grain size and composition throughout the component. The heating is carried out above critical temperature followed by cooling in open air. It differs from annealing in that the component is heated to a higher temperature and then removed from the furnace for air cooling. Normalizing refines grain structure and removes internal stresses induced during the operations such as welding, casting, forging, forming or machining. Stress, if not controlled, results to failure of metal. Hence, before hardening steel, it should be normalized to ensure desired results. Usually, low-carbon steels do not require normalizing. However, if these steels are normalized, there is no harm. Castings are generally annealed, rather than normalized.

Tempering

The purpose of tempering is to relieve the stress and brittleness. After hardening, when steel is quenched, it becomes brittle and hard. Also, it contains internal stress. Tempering is the process of reheating the steel to relieve stress and improve mechanical properties. The tempering temperature and time are generally controlled to affect the characteristics of the steel. Tempering consists of heating steel below the lower critical temperature. Higher tempering temperatures are required to impart further ductility, although some yield strength is lost. This process may also be carried out on normalized steels.

11.6 NON-FERROUS METALS AND ALLOYS

Some of the important non-ferrous metals and alloys are discussed below.

11.6.1 Copper

Copper is one of the most widely used non-ferrous metals. It is obtained from its ores, which are oxides, sulphides and carbonated by dry, wet or electrometallurgical process. It is soft, highly ductile and malleable yet very tough. It is not affected by dry air at ordinary temperature but in moist air and in the presence of carbon dioxide, it is slowly converted green basic carbonate. It is a good conductor of heat and electricity. It is used for making coins, electric wires, pipes, heater tubes, rivets and many household items, etc. It can be forged, rolled and drawn into wires. It forms important alloys like brass and bronze.

11.6.2 Copper Alloys

The copper alloys are of two types:

(*a*) Copper-zinc alloy (Brass)

(*b*) Copper-tin alloy (Bronze)

1. **Brass:** Brass is very popular alloy consisting of copper and zinc. It is highly resistant to corrosion. There are different types of brasses, depending upon the proportions of alloying elements. The properties of a brass vary with the percentage of zinc in it. Proportion of zinc in brass may vary from 5 to 45%. With increase in the percentage of zinc, the ductility of the brass increases but after 37%, ductility starts to fall.

   Red brass contains less than 20% zinc. It is used for making plumbing pipes, condenser tubes, hardwares, etc. Cartridge brass contains 28 to 38% zinc. It is more ductile and is used for cold rolled sheets, wires, tubes, etc. Muntz metal (Yellow brass) contains 40% zinc. It has greater tensile strength and lesser ductility. It is suitable for hot working by rolling, extrusion, stamping, etc. It is used for making welding rods, valve stems, etc.

2. **Bronze:** Bronze contains copper and tin in varying proportions. The range of composition is 5 to 25% tin and 75 to 95% copper. It is quite hard, resists surface wear and can be very easily shaped or rolled into wire, rods and sheets. Bronzes are superior to brasses with respect to corrosion resistant properties. It is used in hydraulic fittings, pump linings, in making household utensils, bearings, etc.

   Phosphorus bronze contains 93.7% copper, 6% tin and 0.3% phosphorus. It is used for making bearing, springs, gears, etc. Manganese bronze contains 60% copper, 35% zinc and 5% manganese. It is stronger than phosphorus bronze. It is used for making bushes, plungers, etc. Aluminium bronze contains 88% copper, 8% aluminium, 3% iron and 0.5% tin. It is used for making propellers, components of pump, etc.

11.6.3 Aluminium

Aluminium is one of the most abundant metals in the earth. However, it is never found free. It is obtained from its ore called bauxite, which is hydrated aluminium ore. The extraction of aluminium from bauxite consists of following two steps:

1. Purification of bauxite, and
2. Electrolysis of purified ore to obtain aluminium.

In its pure state, the metal is soft and weak, but when mixed with other alloys, it becomes hard and rigid. Aluminium is a soft, durable, light in weight, ductile and malleable metal. It can be easily formed, machined, casted, drawn and extruded. It is a good conductor of electricity and heat. It is used for making overhead cables. It is non-magnetic and does not easily ignite. Corrosion resistance of aluminium is excellent due to thin surface layer of aluminium oxide. This layer is formed when the metal is exposed to air, which prevents further oxidation.

It is extensively used in aircraft and automobile components. It is also used for making cooking utensils and thin foils (for wrapping food items).

11.6.4 Aluminium Alloys

Aluminium alloys are those in which aluminium is the predominant metal alongwith other alloying elements such as copper, magnesium, manganese, nickel, silicon and zinc. One element can be either used alone or combination of two or more may be used. Some of the important aluminium alloys are discussed below.

1. **Duralumin:** It contains 94% aluminium, 4% copper, 0.5% manganese, 0.5% magnesium, 0.5% iron and 0.5% silicon. It has high tensile strength and electrical conductance. It is light in weight. It is used in automobile and aircraft applications. It can be forged, casted and stamped. It is also used in manufacturing of connection rods, bars, bolts, rivets, etc.
2. **Y-alloys:** It is also called as copper-aluminium alloy. The addition of copper to aluminium improves its strength and machinablity. Y-alloys contain 92.5% aluminium, 4% copper, 1.5% manganese, 2% nickel and small amount of silicon, magnesium and iron. It is used for making cylinder heads, crankcase and pistons of I.C. engine.
3. **Magnalium:** It contains aluminium, magnesium and copper. It is made by melting the aluminium with 2-10% magnesium in a vacuum and then cooling it in a vacuum. It contains 1.75% copper. It is brittle and light in weight. It has good machinablity. It is used for making aircraft components.
4. **Hindalium:** It is an alloy of aluminium and magnesium with small quantities of chromium and silicon. It is strong and hard. It can't be scratched easily. It is used for manufacture of utensils.

11.6.5 Magnesium

Magnesium is having low density, which means it can be used for making very light weight structures and aircraft components. It is very important property of magnesium. Also, it has good corrosion resistance, less ductility and moderate strength. Powered magnesium ignites readily and burns with white flame. However, magnesium in the form of castings, sheet, etc. does not present any fire risk. Pure magnesium finds very little application and it is the alloys of magnesium which are used for industrial purposes.

11.6.6 Zinc

At low temperature, it is soft and brittle but at high temperature, it is malleable. Zinc can be rolled into thin sheets. It is corrosion resistant and is used as a protective coating on iron in the form of galvanized surface. It is also used for making plates in batteries, fastening device, printing blocks, etc. It is also used for making different alloys.

11.6.7 Tin

It is brightly shining metal. It is soft, malleable and ductile. It can be rolled into very thin sheets. It is used for making different alloys like solder, as a protective coating for iron and steel sheets and for making tin foil used as moisture proof packing.

11.6.8 Nickel and its alloys

Nickel is silvery white metal. It is corrosion resistant, hard and ductile. Its melting point is 1450°C. It is used as alloying element in different types of steel. It is also used as protective covering on iron and brass.

When 67% nickel is alloyed with 30% copper and 3% aluminium, it produces an alloy called as Monel metal. It has high tensile strength and good resistance to corrosion. Monel metal is used in chemical and food-handling industries. It is also used for making turbine blades.

Another alloy of nickel is nichrome, which contains 60% nickel, 15% chromium and 20% iron. Nichrome is non-corrosive and can withstand high temperature. It is used for making electrical resistors and heating elements for heaters.

11.7 NON-METALLIC MATERIAL

There are number of non-metallic material used in engineering applications. Some of these are described below.

11.7.1 Ceramics

The word ceramic comes the from Greek word "*keramikos*", which means pottery. Ceramics are inorganic, non-metallic elements that are manufactured from powdered materials and are fabricated into products through the application of heat. They display characteristic properties such as hardness, strength and brittleness. They have excellent resistance to oxidation and corrosion. Ceramics can also be used as electrically conductive materials or insulators. They have high resistance to high temperature and harsh environment.

Ceramics are typically crystalline in nature and are compounds formed between metallic and non-metallic elements such as aluminium and oxygen (alumina-Al_2O_3), calcium and oxygen (calcia-CaO) and silicon and nitrogen (silicon nitride-Si_3N_4). Depending on their method of formation, ceramics can be dense or light in weight. Ceramic finds numerous applications, some of the categories that make up the ceramic industry are:

(*a*) Structural clay products (such as brick, sewer pipe, roof and wall tile, flue linings, etc.)

(*b*) Whitewares (such as dinnerware, floor and wall tile, electrical porcelain, etc.)

(*c*) Refractories (such as brick and monolithic products used in metal, glass, cements)

(*d*) Glasses (flat glass such as windows, container glass such as bottles, pressed and blown glass such as dinnerware and advanced glass such as optical fibers)

(*e*) Abrasives (natural such as diamond) and synthetic (such as silicon carbide, fused alumina). Abrasives are used for grinding, cutting, polishing, lapping, or pressure blasting.

(*f*) Cements

(*g*) Advanced ceramics

- Structural (bioceramics and cutting tools)
- Electrical (insulators and superconductors)
- Coatings (engine components)

The atoms in ceramic materials are held together by a chemical bond. The two most common chemical bonds for ceramic materials are covalent and ionic. Covalent and ionic bonds are much stronger in ceramic as compared to metallic bonds. Hence, ceramics are brittle and metals are ductile.

11.7.2 Polymers

The word polymer literally means many parts. Polymers are the material that contains many chemically bonded parts, which are bonded together to form a solid. Polymers have many properties that make them attractive. Some of the important properties are:

- They are light in weight.
- They are less dense than metals.
- They resist harsh environment, decay and corrosion.
- They can be easily fabricated.
- They have got good surface finish.
- They offer good compatibility with human tissue.

Two of the most important polymers are plastics and elastomers. Plastics are a large and varied group of synthetic materials, which are processed by forming or moulding into shape. There are many types of plastics such as polyethylene and nylon. Elastomers (or rubbers) are polymeric materials having high elasticity. They can be elastically deformed when force is applied to them and return to their original position when the force is released.

Animal fibre, natural rubber, leather, silk are examples of natural occurring polymer. However, most polymers are manufactured by the combination of hydrogen, oxygen, carbon atoms and other non-metallic substances. Such types of polymers are known as synthetic polymer. Example are plastic, rubber, lubricants and solvents, etc.

Depending on how polymer is structurally and chemically bonded, they can be divided into two groups: thermoplastics and thermosetting plastics. Thermoplastics are those plastics which can be softened by heating and moulded into shape again. They include polyethylene, polypropylene, polystyrene and polyvinyl chloride. Thermosetting plastics cannot be softened and reworked. Thermosetting plastics include alkyds, amino and phenolic resins, Bakelite, epoxies and unsaturated polyesters, etc.

11.7.3 Composites

A composite can be defined as a combination of two or more different materials, each of which retains its own distinctive characteristics to create a new material with properties that cannot be achieved by any of the materials alone. Wide ranges of engineering materials come in the category of composite. For example, concrete is a composite because it is a mixture of cement and aggregate.

Composite materials have two phases. The reinforcing phase is the fibers, sheets or particles that are embedded in the matrix phase. The reinforcing material and the matrix material can be metal, ceramic or polymer. Reinforcing materials are strong with low densities while the matrix is usually a ductile or tough material. Some of the common types of composites are reinforced plastics, metal-matrix composites, ceramic-matrix composites, sandwich structures and concrete.

Composite materials can take various forms but they can be divided into three main categories depending on the strengthening mechanism. These categories are dispersion strengthened, particle reinforced and fiber-reinforced. Dispersion strengthened composites have a fine distribution of secondary particles in the matrix of the material. Particle reinforced composites have a large volume fraction of particle dispersed in the matrix and the load is shared by the particles and the matrix. In fiber-reinforced composites, the fiber is the primary load-bearing component.

If the composite is designed and fabricated correctly, it combines the strength of the reinforcement with the toughness of the matrix to achieve a combination of desirable properties not available in any single material. Some composites also have the merit of being tailorable so that properties, such as strength and stiffness, can easily be changed by changing amount or orientation of the reinforcement material.

SUMMARY

- Corrosion is the deterioration of materials by chemical interaction with their environment.
- The rate of corrosion is expressed as the loss in weight of specimen per unit surface area per unit time.
- The environmental conditions, under which an object is maintained and operated has a great affect on the corrosion characteristics.
- Corrosion characteristics are affected by foreign material which sticks to the surface of the metal like dust, grease etc.
- Galvanic corrosion occurs when two different metals are placed in contact with each other.
- Pitting Corrosion occurs due to rupture of protective film on a surface of the metal which may take place either by surface irregularities or by local straining.
- Stress Corrosion occurs when ductile metals are subjected to a tensile stress in a corrosive environment, especially at elevated temperature.
- Erosion Corrosion occurs as the result of a combination of an aggressive chemical environment and high fluid-surface velocities.
- Hydrogen damage is metal deterioration processes due to interaction with hydrogen.
- Electroplating is the process in which metals are deposited from solutions containing metallic ions by means of electrolysis.
- In galvanizing process, the object is cleaned thoroughly and dipped into a bath of molten zinc. It is to provide zinc coating on the object.
- Fine plastic powder or liquid PVC can be coated to the materials using this method. The hot metal piece is to be dipped into the bath to produce the required coating. Plastic coatings are widely used in pumps, vessels, valves, and pipelines for process industries. They can be used in food industries, oil industries and also in water treatment industries.
- Phosphate coating is used on steel for corrosion resistance, lubricity or as a foundation for subsequent coatings.
- Metal spraying is the method of coating metal on the surface of objects. It is done by feeding a metal wire at a controlled rate into a gun.
- Paints are used not only to improve the appearance of the object but also, they protect it from corrosion.
- Parkerizing consists of producing a film of phosphate on the metal by immersing the metal into a bath of phosphoric acid.
- Enamelling is a protective coating of fused silicate on the surface of metal.
- A silvery-white, frosted appearance can be given to aluminium by applying a caustic solution to its surface in a process called as chemical etching.
- Bright dipping is a special type of chemical finishing process that yields a bright, mirror-like finish called as specular finish.

ASSIGNMENT

1. Students should be asked to write literature review of nano-materials. Literature review must include the chemical composition of the nano- materials, their production methods, characteristics, structure properties and their applications.

EXERCISES

1. How are engineering materials classified? On what basis, is material selected for a particular job?
2. Differentiate between ferrous and non-ferrous materials.
3. What is meant by mechanical property of engineering material? Describe any five mechanical properties.
4. Explain briefly the following properties of engineering materials: strength, toughness, hardness and hot hardness.
5. Differentiate between ductility and brittleness.
6. Define the following properties of materials: stiffness, elasticity, plasticity, resilience and creep.
7. What is meant by thermal expansion?
8. Differentiate between thermal and electrical conductivity.
9. What do you understand by magnetic permeability?
10. Distinguish clearly between pig iron and wrought iron. Also, mention their applications.
11. How is cast iron obtained? Classify and explain different types of cast iron. Give examples of their applications.
12. Differentiate between grey and white cast iron.
13. What is meant by chilled cast iron?
14. Define plain carbon steel. Write notes on different type of carbon steel. Give examples of their applications.
15. What is meant by alloy steel? What is the effect of addition of different elements to steel?
16. Write notes on stainless steel and high speed steel.
17. Write short notes on free cutting steel.
18. What is alloy tool steel? What are its different types? What is the composition of 18-4-1 HSS?
19. What is meant by heat treatment of steel? Write notes on hardening and annealing.
20. Explain the case hardening process.
21. What do you understand by carburizing? Explain in detail.
22. Describe the process of cyaniding and nitriding.
23. What is meant by flame hardening and induction hardening?
24. Explain the technique of normalizing and tempering.
25. Discuss copper and its alloys. Mention their applications also.
26. Enumerate the properties of aluminium. Give properties and uses of different alloys of aluminium.
27. Describe briefly different non-metallic materials used in engineering practices.
28. Write notes on polymers.
29. What is meant by composites? Describe in detail.
30. What do you understand by ceramics? Classify ceramic materials.

❑❑❑

Corrosion–Its Control and Chemical Cleaning

12
CHAPTER

12.1 INTRODUCTION

Corrosion is the deterioration of materials by chemical interaction with their environment. There are numerous examples of corrosion in day to day life. The term corrosion can also be applied to the degradation of plastics, concrete and wood but usually it is used for metals. The most widely used metal is iron (steel).

The rate of corrosion is expressed as the loss in weight of specimen per unit surface area per unit time. The yearly economic loss due to corrosion is quite high. The metals worth trillions of rupees are being wasted because of corrosion.

The consequences of corrosion are many and they can effect operation of equipment which is more serious than the loss of material. Failures of various types and the need for replacements may be required even though less amount of metal is destroyed.

Some of the ill effects of corrosion are described below:

(*i*) Reduction of thickness of metal which can result in loss of mechanical strength and structural failure. When the metal loss in localized zones is such that it gives crack like structure, considerable weakening may take place from very small amount of metal loss.

(*ii*) Injuries to people can arise from structural failure (such as bridges, and automobile components).

(*iii*) Mechanical damage to valves, pumps, etc., or blockage of pipes by solid corrosion products can take place.

(*iv*) Value of goods is reduced because of deterioration of appearance.

(*v*) Loss of technically important surface properties of a metallic component may take place which includes frictional and bearing properties.

(*vi*) Perforation of vessels and pipes which can allow escape of their contents and possible harm to environment.

(*vii*) Corrosion can add complexity in design and increase cost of equipment which needs to be designed to withstand corrosion.

(*viii*) It also results in loss of efficiency.

Corrosion involves the chemical reaction of a metallic material with surrounding environment. It is a natural process in the sense that the metal is attempting to revert to the chemically combined state in which it is found in nature. Hence, it is a process that may be expected to occur and should not be considered as inevitable. Control or prevention of corrosion is possible through different methods.

12.2 FACTORS AFFECTING CORROSION

There are various causes of corrosion which can affect its type, speed, cause and seriousness. Though some of these factors can be controlled, they are others which cannot be controlled. Various factors affecting corrosion are discussed below.

(*i*) **Climate:** The environmental conditions under which an object is maintained and operated, significantly affect the corrosion characteristics. Exposure to water and moisture-laden air is detrimental than dry climate.

(*ii*) **Temperature:** Temperature considerations are also very important as speed of electrochemical attack is increased in a hot climate.

(*iii*) **Size and Type of Metal:** It is an established fact that some metals corrode faster than others, but it is a less known that variations in size and shape of metal can also affect the corrosion characteristics. Thick structural sections are more susceptible to corrosion as compared to thin sections.

(*iv*) **Foreign Material:** Among the controllable factors which affect corrosion characteristics are foreign material which sticks to the surface of the metal. Such foreign material includes the following:

(*a*) Atmospheric dust

(*b*) Oil grease and exhaust residues

(*c*) Salt water and moisture condensation

(*d*) Spilled battery acids

(*e*) Welding and brazing flux residues.

12.3 TYPES OF CORROSION

Corrosion is classified as below.

(*i*) **Nature of corrodent:** Corrosion can be classified as wet or dry. Dry corrosion involves reaction with gases at elevated temperature. Wet corrosion takes place in presence of water or a conducting liquid.

(*ii*) **Mechanism of Corrosion:** This can involve direct chemical attack or indirect electro-chemical attack.

(*iii*) **Appearance of the corroded metal:** Corrosion can be either uniform or localized. In case of uniform corrosion, the entire surface of the metal gets evenly corroded. On the other hand in case of localized corrosion, only small areas are affected.

12.4 SPECIFIC TYPES OF CORROSION

Some important types of corrosion are described below:

12.4.1 Galvanic Corrosion

This type of corrosion occurs when two different metals are placed in contact with each other. It is caused by the willingness of one to give up electrons as compared to other. Three important features need to operate for corrosion to take place. The metals need to be in contact with each other electrically. One of the metals needs to be significantly better at giving up electrons than the other. An additional path for ion and electron movement is required.

In order to prevent galvanic corrosion, it should be ensured that one or more of the above features do not exist. Electrical contact of metals should be broken with plastic insulators or coatings. Metals selected should be close to each other in the galvanic series. Ion movement should be prevented by coating the junction with an impermeable layer or ensure environment is dry and liquids cannot be trapped.

12.4.2 Pitting Corrosion

This type of corrosion occurs due to rupture of protective film on a surface of the metal. This rupture may take place either by surface irregularities or by local straining. It can also occur by the action of foreign materials. The exposed metal gives up electrons quite easily and the reaction initiates tiny pits with localized chemistry in supporting rapid attack. Control of pitting can be ensured by:

(*i*) Maintaining the protective layer of the material.
(*ii*) Selecting a resistant material.
(*iii*) Control of the chemistry of fluids
(*iv*) Use of inhibitors.
(*v*) Providing protective coating.

Pitting corrosion is very common with stainless steel and aluminium in aqueous environment containing metallic chloride salts.

12.4.3 Selective Attack

This type of corrosion occurs in alloys (like brass) when one component is more susceptible to attack than another and corrodes preferentially leaving a porous material that crumbles. It can be avoided by selecting a resistant material.

Some other means that can be effective employed for control of this type of corrosion are:

(*i*) Coating the material
(*ii*) Reducing the aggressiveness of the environment
(*iii*) Use of cathodic protection

12.4.4 Intergranular Corrosion

This type of corrosion occurs mostly on grain boundaries of the crystals that form the metal. Identification of intergranular corrosion usually requires microstructure examination. This type of corrosion results from local differences in composition, like coring commonly encountered in alloy castings. Grain boundary precipitation, such as chromium carbides in stainless steels, is a well recognized and accepted mechanism of intergranular corrosion. The precipitation of chromium carbides consumes the alloying element chromium from a narrow band along the grain boundary and this makes the zone anodic to the unaffected grains. The chromium depleted zone becomes the preferential path for corrosion, if under tensile stress.

12.4.5 Stress Corrosion

This type of corrosion occurs when ductile metals are subjected to a tensile stress in a corrosive environment, especially at elevated temperature. This is specific to a metal combined with a specific environment. The environment that causes stress corrosion for a given metal can be only mildly corrosive to the other metal. It can be prevented by reducing overall stress level and designing out stress concentrations. Metals selected should not be susceptible to the environment. Use of protective coatings is also effective protection technique.

12.4.6 Erosion Corrosion

This type of corrosion occurs as the result of a combination of an aggressive chemical environment and high fluid-surface velocities. This can be due to fast fluid flow past a stationary object, like the oil-field check valve or from the quick motion of an object in a stationary fluid, for example when a ship's propeller churns the ocean. Surfaces which have undergone erosion corrosion are usually clean unlike the surfaces from other types of corrosion. It can be controlled by the using harder alloys or by using corrosion resistant alloy. Changes in fluid velocity and changes in flow patterns can also significantly reduce its effects.

12.4.7 Hydrogen Damage

Hydrogen damage is the name given to a number of metal deterioration processes due to interaction with hydrogen. Hydrogen can cause significant corrosion problems. Hydrogen embrittlement is a problem with high-strength steels, titanium, and some other metals. It can be controlled by eliminating hydrogen from the environment or by the using resistant alloys.

Hydrogen blistering can take place when hydrogen enters steel due to reduction reaction on a metal cathode. Single-atom nacent hydrogen atoms then diffuse through the metal until they meet with another atom, generally at inclusions or defects in the metal. The resultant diatomic hydrogen molecules are then too big to migrate. Finally, a gas blister builds up and may split the metal.

12.4.8 Microbial Corrosion

This type of corrosion occurs by the action of bacteria or fungi on material, which can lead to degradation of materials. It can take place by various actions like attack of the metal or protective coating by acid by-products, sulphur, hydrogen sulphide or ammonia. It can also occur by direct interaction between the microbes and metal. It can be prevented by frequent cleaning, selection of resistant materials, use of biocides and cathodic protection.

12.5 CORROSION MINIMIZATION AND PROTECTION

The corrosion may be minimized or protected by different methods, some of which are described below.

12.5.1 Electroplating

Electroplating is defined as a process in which metals are deposited from solutions containing metallic ions by means of electrolysis. It is extremely useful method of corrosion protection and can also be applied for decoration purposes. It changes the chemical, physical, and mechanical properties of the object being electroplated.

The process used in electroplating is called electro-deposition. It is similar to a galvanic cell but acting in reverse direction. Electroplating consists of setting-up the object to be coated as cathodes in electrolytic bath, the anode being metallic zinc or nickel. When current is passed through electrolyte, the metal ions get deposited on the object. A power source supplies a direct current to the anode, oxidizing the metal atoms that comprise it and allowing them to dissolve in the electrolyte. At the cathode, the dissolved metal ions in the electrolyte are reduced at the interface between the solution and the cathode, in such a way that they are plated onto the cathode. The thickness of coating is a function of current density and time. Generally, zinc, tin, nickel, chromium coating are applied by this method.

12.5.2 Galvanizing

Galvanizing is a hot coating process. The object is cleaned thoroughly and dipped into a bath of molten zinc. It is to provide zinc coating on steel products. Sometimes, the objects may be placed within zinc dust in a drum and is heated upto 300°C. This method is extremely useful for small object like bolts, pipe fittings, etc.

Zinc coatings prevent corrosion by forming physical barrier and by acting as a sacrificial anode, in case the barrier gets damaged. When exposed to the atmosphere, zinc reacts with oxygen to form zinc oxide, which further reacts with water in the air to form zinc hydroxide. This zinc hydroxide reacts with carbon dioxide in the atmosphere to yield a thin, impermeable, tenacious and insoluble dull gray layer of zinc carbonate which adheres to the zinc. This layer protects zinc from corrosion, in a way similar to the protection afforded to aluminium and stainless steels by their oxide layers.

Galvanizing can be applied to number of items such as boilers, pipes, tanks, cans and exposed steel structures. Tin and lead coatings are also produced using this method.

12.5.3 Aluminium Coating

Aluminium coating is an economical process for inhibiting such corrosion by protecting the surface of metals operating in severe high temperature. It cannot be deposited by electro-deposition. Examples are aluminium windows, furniture, elevator doors, decks, etc. In this method, a rotating drum is used. The drum contains aluminium powder, aluminium oxide and chlorine. The object to be coated is placed in the rotating drum. Such coatings applied to duralumin are extremely successful. But when applied to steel, the layers of such coatings are brittle in nature.

12.5.4 Thermo-plastic Coating

Fine plastic powder or liquid PVC can be coated to the materials using this method. The hot metal piece is to be dipped into the bath to produce the required coating. Plastic coatings are widely used in pumps, vessels, valves, and pipelines for process industries. They can be used in food industries, oil industries and also in water treatment industries.

12.5.5 Phosphate Coating

Phosphate coating is used on steel for corrosion resistance, lubricity or as a foundation for subsequent coatings. It serves as a conversion coating, in which dilute solution of phosphoric acid and phosphate salts are applied by spraying or immersion and chemically reacts with the surface of the object being coated to form a layer of insoluble, crystalline phosphates. Phosphate coatings can also be used on aluminium, zinc, cadmium, silver and tin. They are also used in anti-rust coating.

12.5.6 Metal Spraying

Metal spraying is the method of coating metal on the surface of objects. It is done by feeding a metal wire at a controlled rate into a gun. Metal spraying offers numerous advantages over other methods like a greater working speed, applicability to large surfaces and ease of application of restricted areas. Two, three or four coats are usually applied to get final desired thickness. Sprayed molybdenum is used on aluminium pulley, plug gauges, crankshafts, etc.

12.5.7 Painting

Paints are used not only to improve the appearance of the object but also, they protect it from corrosion. Painting is the practice of applying paint, enamels, pigment, or other medium to the object. The application of the medium is commonly applied with a brush but other methods can also be used. Painting is used in external parts of automobile, steel bridges, various household items, etc. which are exposed to the atmosphere. Painting provides resistance to corrosion against ordinary atmosphere.

12.5.8 Salt and Oxide Film

Salts and oxide films are obtained, by reacting metal with a solution, which produces required film. These films prevent corrosion. Some of examples are described below.

Chromate Pickle

Chromate coating is a type of coating used to passivate aluminum, zinc, cadmium, copper, silver, magnesium and tin alloys. It is used as a corrosion inhibitor, primer and for decorative purposes. It consists of immersing of object into a bath of nitric acid and dichromate. A thin layer is formed on the object, which protects it from corrosion. Chromate coatings can be applied on everyday items like hardware and tools.

Parkerizing

Parkerizing is a method of protecting steel from corrosion and increasing its wear resistance. It consists of producing a film of phosphate on the metal by immersing the metal into a bath of phosphoric acid. It can be viewed as an improved zinc or manganese phosphating process. Parkerizing is commonly used on firearms. The Parkerizing process cannot be used on non-ferrous metals like aluminum or copper. However, it cannot be applied to steels containing nickel or on stainless steel itself.

Enamelling

Enamelling is a protective coating of fused silicate on the surface of metal. The word enamel comes from the German word *"smelzan"*. *Enamel means* a coloured glassy substance, transparent or opaque, fused to the surface of object. When these glass particles are heated in a furnace at about 800°C on a red hot metal, this result in a beautiful coloured glass coating on the surface of metal. This coating is very brittle and can be easily damaged. It is used for protecting steel, cast iron and other metals from corrosion.

Slushing

When machine parts are to be transported, a layer of grease is applied to them. The method of applying grease is called as slushing. Also, when machine is to be stored for sometime, a thick layer of grease is applied to various parts of machine. This prevents corrosion of machine parts. Some additives may be added to grease to increase its effectiveness.

12.6 CHEMICAL CLEANING

It is generally accepted that the effectivenss of a coating can be improved by good surface preparation before application. Surfaces must be free from impurities that may be corrosive or may impair adhesion or continuity of coating or restrict access of inhibiting vapours. These impurities may be comprised of debris from previous processing, mineral oils, corrosion inhibitors, etc. Chemical cleaning means removal of such impurities from the surface of the metal.

The process of cleaning should be powerful enough to remove the undesirable materials from the surface to be cleaned. The impurities can dictate the choice of cleaning. Solvent degreasing, an alkali wash or steam cleaning would remove oil or grease. Mechanical, acid and electrolytic techniques can be used to remove scale or rust. In some cases, a simple water wash can be effectively employed but others may require a combination of more than one treatment. Manufacturers give their recommended method of removal on the technical data sheet for each product.

Cleaning activities do not have any well-defined classification. However, they can be broadly classified with respect to type of film to be cleaned. Such as:

(*a*) Removal of organic films

(*b*) Removal of inorganic films

12.6.1 Removal of Organic Films

Organic films can be divided into different groups like mineral, animal or plant oils. Mineral oils are neither soluble in water nor in acid or in alkaline solution, therefore they are cleaned by using chlorinated hydrocarbons. For the cleaning purpose, naptha (petroleum solvents) and tri-choloroethylene are used. It also removes fats and greases. Animal oils get accumulated on the surface of machine from lubricating oils. They are also soluble in chlorinated hydrocarbons. In some cases, emulsifying solvents can also be used. They contain both penetration solvent and dissolved emulsifying agents. Alkaline cleaning is old method of removing vegetable oils. Alkaline cleaning also removes greases, marking pens, fingerprints, or shop dirt. Caustic soda is commonly used.

12.6.2 Removal of Inorganic Films

Inorganic films may include scale, smut, rust, tarnish, dirt or cleaning residues. The removal of scales and smut is a very difficult task. Sometimes, mechanical techniques are used to remove them. Acid cleaning is used to remove inorganic contaminants like oxide films and intermetallics, which might interfere with finish.

Rust and tarnish are formed under the influence of corrosive environment. These films react readily with acid and cynide. Inhibited muriatic acid or sulphuric acid can be used effectively for dissolving iron oxides and calcium containing scales. For stainless steel nitric acid can be used. The term 'dirt' is a miscellaneous title that includes all types of inorganic residues. They can be cleaned by hand or machine and brushing or by some other mechanical technique.

Mechanical techniques include hydraulic spraying, steam spraying, polishing and other abrasive techniques. Buffing is another mechanical method which provides lustrous finish to the surface.

12.7 FINISHES

Different types of finishes are described below.

12.7.1 Mechanical Finishes

The surface of aluminium can be buffed and burnished to a mirror finish and blasted to a rough texture. Blasting include abrasive blasting, shot blasting, and glass-bead blasting. Other mechanical finishing methods include sanding, polishing, and tumbling. These methods can be applied as a final surface finish, to improve surface quality or in preparation for a final cosmetic finish.

12.7.2 Chemical Finishes

Etching

A silvery-white, frosted appearance can be given to aluminium by applying a caustic solution to its surface in a process called as chemical etching. The aluminum is passed through a hot bath, rinsed, and then immersed in a de-oxidizedesmut bath that removes undissolved alloy constituents or impurities. Further rinses are usually required to complete the process.

Bright Dipping

Bright dipping is a special type of chemical finishing process that yields a bright, mirror-like finish called as specular finish. The bath usually contains both phosphoric acid and nitric acid, heated to high temperature. The object to be bright-dipped must be first polished to remove fine scratches, and afterwards is usually anodized. Anodizing the bright-dipped part serves dual purposes: it protects the surface finish, and can be used to apply color to the surface.

SUMMARY

- The metals, which have iron as their main constituent, are called as ferrous metals e.g. cast iron, steel etc.
- The metals which have elements other than iron as their main constituent are called as non-ferrous metals like copper, aluminium, magnesium, zinc etc.
- Mechanical properties indicate material reaction to the application of mechanical forces and load.
- Stiffness is the property of material by virtue of which it resists deformation under stress. It is also known as rigidity.
- Hardness is the property of material by virtue of which it resists abrasion and penetration.
- Creep is a slow, time-dependent, progressive deformation under constant loading of the material.
- When two dissimilar metals are joined from end to end and their junction is maintained at different temperatures, a small voltage is produced and this is known as thermoelectric effect.
- Magnetic permeability is defined as the ease with which a material can be magnetized.
- Cast iron contains 2.5 to 5% carbon alongwith varying portions of silicon, sulphur, phosphorus and manganese.
- Cast iron being brittle cannot be used in those parts, which are subjected to shocks.
- Mild carbon steel contains carbon from 0.05 to 0.3%.
- High carbon steel contains carbon from 0.6 to 1.5%.
- The important alloying elements of alloy steel are nickel, chromium, manganese, silicon, tungsten and vanadium etc.
- The popular 18-4-1 HSS contains 18% tungsten, 4% chromium and 1% vanadium.
- Case hardening is a diffusion process in which an alloying element (carbon or nitrogen) is diffused into the surface of metal.
- Flame hardening is a form of localized hardening. This process involves use of oxy-acetylene flame on the surface of the steel to be hardened.
- Brass is very popular alloy consisting of copper and zinc. Proportion of zinc in brass may vary from 5 to 45%.
- Bronze contains copper and tin in varying proportions. The range of composition is 5 to 25% tin and 75 to 95% copper.
- Ceramics are inorganic, non-metallic elements that are manufactured from powdered materials and are fabricated into products through the application of heat.

- Polymers are the material that contains many chemically bonded parts, which are bonded together to form a solid.
- A composite can be defined as a combination of two or more different materials, each of which retains its own distinctive characteristics to create a new material.

ASSIGNMENT

1. Students should be shown different corrosion protection techniques.

EXERCISES

1. What is meant by corrosion? Explain its mechanism. Also, write down harmful effects of corrosion.
2. Discuss various factors that affect corrosion.
3. What are different types of corrosion?
4. Differentiate between wet and dry corrosion.
5. Explain the mechanism of galvanic corrosion. How it can be prevented?
6. Write informative notes on pitting corrosion.
7. Explain the mechanism of intergranular corrosion. How it can be prevented?
8. What do you understand by selective attack? Explain.
9. Explain mechanism of stress corrosion and erosion corrosion.
10. Write short notes on hydrogen damage.
11. Describe briefly microbial corrosion.
12. Describe different methods of corrosion protection.
13. What is meant by electroplating? How is it done?
14. What do you understand by galvanizing? How is it done?
15. Explain aluminium coating process.
16. Write short notes on
    (*a*) Thermo-plastic coating (*b*) Phosphate coating
    (*c*) Metal spraying
17. Differentiate between painting and enamelling.
18. Discuss the following:
    (*a*) Chromate pickle (*b*) Parkerizing
    (*c*) Enamelling (*d*) Slushing
19. What is chemical cleaning? In what cases is it employed?
20. Describe different types of chemical cleaning.
21. Describe different types of finishes.
22. Explain difference between etching and bright dipping.

❑❑❑

Objective Questions

CHAPTER 1: SAFETY PRACTICES

1. .... is mainly concerned with minimizing, controlling and eliminating hazards.
   (*a*) Industrial intelligence (*b*) Industrial automation
   (*c*) Industrial safety (*d*) All of the above
2. Safety consciousness is mainly concerned with development of awareness in
   (*a*) Customers (*b*) Workers
   (*c*) Both (*d*) None of these
3. Safety consciousness can be increased by adopting the given methods:
   (*a*) Forming safety committees.
   (*b*) Organizing safety awareness programs for the workers.
   (*c*) Training workers to use safety devices.
   (*d*) All of the above.
4. Causes of accidents in industries are
   (*a*) Obsolete or poorly maintained equipment
   (*b*) Unguarded and untested machines
   (*c*) Violation of norms (*d*) All of the above
5. The device used for either lifting or lowering the loads is
   (*a*) Hoist (*b*) Lathe machine
   (*c*) Portable elevator (*d*) All of the above
6. The device used for moving the material either vertically or horizontally between fixed stations is
   (*a*) Lathe machine (*b*) Conveyor
   (*c*) Portable elevator (*d*) None of these
7. Important precaution related to handling of hoisting equipments is
   (*a*) It should be of proper mechanical construction and free from defects.
   (*b*) It should be operated by fully trained operator.

(*c*) Repair and maintenance work should be carried out only when it is switched off.

(*d*) All of these.

8. Important safety guideline related to handling of conveyors is

(*a*) All rotating parts should be properly guarded.

(*b*) It can be operated at any speed as required.

(*c*) The persons working on or near the conveyor can wear any type of clothes.

(*d*) All of these.

9. Which one of the following safety guideline should be followed while working on drilling machine:

(*a*) Parts of the drill machine should be properly guarded

(*b*) Job should be properly secured before starting drilling operation.

(*c*) The process of drilling must be carried out at proper speed.

(*d*) All of these.

10. Safety precautions to be kept in mind while working on grinder are

(*a*) Flammable substances should be stored at a safe distance from the place where grinding operation is done.

(*b*) Operator working on grinder can wear any type of clothes.

(*c*) Grinder operation should be carried out in air-conditioned room.

(*d*) None of these

CHAPTER 2: SYSTEM OF LIMITS AND FITS

1. The system of manufacture in which dimensions of a part lie within certain specified limits causes

(*a*) Interchangeable manufacture (b) Mass production

(*c*) Quality assurance (*d*) Lean manufacture

2. The size which is obtained by calculation for strength is known as:

(*a*) Actual size (*e*) Allowance

(*c*) Basic size (*d*) Maximum size

3. The dimension as measured on a manufactured part is called as:

(*a*) Actual size (*b*) Allowance

(*c*) Basic size (*d*) Maximum size

4. The maximum size permitted for the part is known as:

(*a*) Low limit (*b*) Limit of size

(*c*) Nominal size (*d*) High limit

5. The minimum size permitted for the part is known:

(*a*) Limit of size (*b*) Design size

(*c*) Low limit (*d*) High limit

6. .... is the difference between maximum and minimum limits of the component.

(*a*) Allowance (*b*) Tolerance

(*c*) Interference (*d*) None of these

7. .... is an intentional difference between the maximum material limit of mating part.
(*a*) Tolerance (*b*) Interference
(*c*) Allowance (*d*) All of the above
8. The algebraic difference between an actual size and corresponding basic size is known as:
(*a*) Fundamental deviation (*b*) Tolerance
(*c*) Allowance (*d*) Actual deviation
9. .... is either the upper or lower deviation which is the nearest one to the zero line for either a hole or shaft.
(*a*) Actual deviation (*b*) Fundamental deviation
(*c*) Tolerance (*d*) Allowance
10. In unilateral system, both the limits of size are on the
(*a*) Same side of the zero line (*b*) Different side of the zero line
(*c*) Zero line (*d*) None of the these
11. In which system, one of the limits of size is on one side of the zero line and the other limit of size is on the other side of the zero line.
(*a*) Unilateral system (*b*) Bilateral system
(*c*) Both (*a*) and (*b*) (*d*) None of the above
12. In which fit, the internal member fits in an external member and always leaves clearance between the parts?
(*a*) Transition fit (*b*) Interference fit
(*c*) Clearance fit (*d*) None of these
13. In which of the following fit, there is zero allowance between mating parts?
(*a*) Snug fit (*b*) Force fit
(*c*) Tight fit (*d*) Wringing fit
14. In .... tolerance zone of shaft and hole intersect.
(*a*) Interference fit (*b*) Transition fit
(*c*) Clearance fit (*d*) None of these
15. In which of the following fit, the internal member is larger than the external member.
(*a*) Clearance fit (*b*) Transition fit
(*c*) Interference fit (*d*) None of these
16. Which of the following fit has maximum negative allowance?
(*a*) Snug fit (*b*) Push fit
(*c*) Both (*a*) and (*b*) (*d*) Heavy force fit
17. A fit between mating parts with slight negative allowance, requiring light to moderate force to assemble is known as:
(*a*) Snug fit (*b*) Tight fit
(*c*) Both (*a*) and (*b*) (*d*) None of these
18. The system in which the limits on the hole are kept constant and limits on the shafts are varied is known as:
(*a*) Hole basis system (*b*) Shaft basis system
(*c*) Hole-shaft system (*d*) Shaft-hole system

19. The system in which the limits on the shaft are kept constant and the variations necessary to obtain the classes of fit are obtained only by varying the limits on the holes is known as:
(*a*) Hole basis system (*b*) Shaft basis system
(*c*) Hole-shaft system (*d*) Shaft-hole system

20. Which of the following fit is used on outer diameter of bus bearing
(*a*) H7/h6 (*b*) M7/m6
(*c*) H7/m6 (*d*) M7/h6

CHAPTER 3: INSTALLATION OF MACHINE TOOLS

1. Alignment and performance tests are collectively known as
(*a*) Acceptance tests (*b*) Final tests
(*c*) Industrial tests (*d*) Practical tests

2. Information related to specifications of machine, its drawings, installation details, test to be performed and maintenance jobs to be carried are given in
(*a*) Foundation plan (*b*) Information manual
(*c*) Erection drawing (*d*) None of these

3. Location of a machine tool means selection of a suitable place where the machine tool can be
(*a*) Overhauled (*b*) Serviced
(*c*) Installed (*d*) Manufactured

4. Foundation of machine tool performs which of the following function
(*a*) It serves as the base to support the entire load of the machine, its mounting, workpiece and tools.
(*b*) It absorbs mechanical vibrations produced during the operation and prevents transmission vibrations from one machine tool to another.
(*c*) It helps to maintain correct alignment of machine tool, its mechanism and moving parts.
(*d*) All of the above

5. .... is a solid foundation of cement concrete or brick masonry, in which, foundation bolts are grouted and the machine rest over the top surface.
(*a*) Block Type (*b*) Box type
(*c*) Frame type (*d*) Wall type

6. .... is hollow box type in construction usually made up of cement concrete and the machine is placed over the top surface.
(*a*) Block Type (*b*) Frame type
(*c*) Box type (*d*) Wall type

7. A horizontal machine tool needs......... foundation as compared to vertical machine tool.
(*a*) lighter (*b*) heavier
(*c*) bigger (*d*) none of these

8. .... foundation of machine tool is required for a higher power motor.
(*a*) Heavier (*b*) Bigger
(*c*) Deeper (*d*) None of these

9. .... shows the profile of the machine and gives particulars as to recesses required to accommodate projections and bolts etc.

(*a*) Foundation plan (*b*) Information manual

(*c*) both (*i*) and (*ii*) (*d*) none of these

10. Which of the following information related to machine foundation is included in foundation plan?

(*a*) Profile and weight of the machine.

(*b*) Shape and size of the foundation to be required.

(*c*) Both (*a*) and (*b*) (*d*) None of these

11. The vibrations caused due to external excitation are known as

(*a*) free vibrations (*b*) forced vibrations

(*c*) external vibrations (*d*) machine vibrations

12. The resistance to the motion of vibratory system is known as

(*a*) damping (*b*) spacing

(*c*) resilience (*d*) none of these

13. Alignment of machine tool should be perfect, as misalignment causes

(*a*) cavitation (*b*) damping

(*c*) vibrations (*d*) none of these

14. Examples of resilient material are

(*a*) cork, linoleum and rubber (*b*) copper and its alloy

(*c*) cement and concrete mixture (*d*) all of the above

15. When the machine is to be protected from transmitting the vibration, it is known as

(*a*) passive isolation (*b*) active isolation

(*c*) defensive isolation (*d*) none of these

16. When the machine is to be protective from vibrations produced from external sources, it is known as

(*a*) passive isolation (*b*) active isolation

(*c*) defensive isolation (*d*) none of these

17. Locating machine away from adjoining structures and establishments is known as

(*a*) geometric isolation (*b*) foundation treatment

(*c*) structural isolation (*d*) defensive isolation

18. Rope pulley block are used for loads

(*a*) upto 1 tons (*b*) from 1 to 5 tons

(*c*) more than 5 tons (*d*) none of these

19. Western pulley block are loads ranging from

(*a*) upto 1 tons (*b*) from 1 to 5 tons

(*c*) more than 5 tons (*d*) none of these

20. Worm block pulley are loads ranging from

(*a*) upto 1 tons (*b*) from 1 to 5 tons

(*c*) more than 5 tons (*d*) none of these

CHAPTER 4: MISCELLANOUS EXAMPLES OF INSTALLATION

1. The flat belts are more .... than the V-belts.
   (*a*) lighter (*b*) heavier
   (*c*) tighter (*d*) versatile
2. If the belt is too tight, it will
   (*a*) put excessive load on bearings (*b*) increase arc of contact
   (*c*) increase coefficient of friction (*d*) decrease arc of contact
3. If the belt is too loose, it may
   (*a*) increase arc of contact (*b*) decrease arc of contact
   (*c*) result in easy slipping (*d*) reduce chances of slipping
4. In a flat belt drive arrangement, belt should run with the slack side on the top in order
   (*a*) to provide decreased arc of contact (*b*) to provide increased arc of contact
   (*c*) to increase coefficient of friction (*d*) to decrease coefficient of friction
5. Pumps instruction manual contains information related to
   (*a*) designing of pump
   (*b*) manufacturing techniques of different components
   (*c*) installation, operation and maintenance of pump
   (*d*) none of these
6. Recommended mass of a concrete foundation for pump should be .... times that of the pump, motor and base.
   (*a*) three (*b*) five
   (*c*) two (*d*) None of these
7. The term "setting" of the boiler generator includes
   (*a*) foundation of the boiler (*b*) only accessories of the boiler
   (*c*) accessories and mountings of the boiler
   (*d*) the material required for surrounding the heating surfaces and the furnace.
8. Boiler must be installed on
   (*a*) carpet (*b*) tiled floor
   (*c*) non-combustible floor (*d*) rubber pads
9. Which of the following functions are performed by the chimney?
   (*a*) it removes flue gases and helps in cleaning the furnace
   (*b*) it removes flue gases and provides draft for the fire
   (*c*) it provides draft and helps in cleaning the furnace
   (*d*) none of these
10. What is the term used to describe the suction that occurs when hot air rises in chimney and creates a vacuum for suction of fresh air into the firebox?
    (*a*) suction pressure (*b*) induction coefficient
    (*c*) draft (*d*) chimney coefficient

11. Which of the following is not the cause of chimney failure?
(*a*) Wind force due to which downward tipping may result.
(*b*) Cavitation
(*c*) Joints failure due to ageing
(*d*) Uneven settlement of foundation.

12. A steam separator should be installed .... the steam turbine.
(*a*) before (*b*) after
(*c*) both (*a*) and (*b*) (*d*) none of these

CHAPTER 5: MACHINE TOOL TESTING

1. Alignment tests deal with
(*a*) dimensional accuracies (*b*) industrial accuracies
(*c*) manufacturing accuracies (*d*) working accuracies

2. Performance tests deal with
(*a*) dimensional accuracies (*b*) working accuracies
(*c*) industrial accuracies (*d*) manufacturing accuracies

3. Both alignment and practical tests are necessary for acceptance of machine tools.
(*a*) True (*b*) False
(*c*) Depends on foundation of the machine (*d*) Depends on type of machine

4. Acceptance tests are conducted on
(*a*) new machines (*b*) reconditioned machines
(*c*) Both (*a*) and (*b*) (*d*) None of these

5. The present day test procedure of alignment test is based on the test chart prepared by
(*a*) Dr. G. Schlesinger (*b*) Dr. Scott Hall
(*c*) Prof. Kevin Nash (*d*) Dr. Bret Nobels

6. Which of the following is included in acceptance tests of machine tools?
(*a*) Tests for operations of different mechanisms.
(*b*) Tests for productive output of the machine tool.
(*c*) Tests for alignment accuracy of the machine tool.
(*d*) All of the above

7. The measuring instruments required for alignment test are
(*a*) dial indicator, spirit level and test mandrel
(*b*) magnetic compass and stop watch
(*c*) dial indicator, stop watch and vernier caliper
(*d*) none of these

8. Which of the following instrument is used for alignment tests to check parallelism and concentricity of holes or rods and flatness of surfaces?
(*a*) Spirit level (*b*) Test mandrel
(*c*) Dial indicator (*d*) Vernier caliper

9. In a dial indicator, 1 division of the dial represents
   (*a*) 0.01 mm travel of the dial indicator (*b*) 0.01 mm travel of the plunger
   (*c*) 1 mm travel of the plunger (*d*) 100 mm travel of the plunger
10. Which of the following instrument is used in alignment tests for checking deflections of long beds in horizontal, vertical or inclined planes?
   (*a*) Dial indicator (*b*) Test mandrel
   (*c*) Autocollimator (*d*) Spirit level
11. Spirit level are specified by
   (*a*) length (*b*) color
   (*c*) weight (*d*) type of spirit used
12. The level of the lathe machine bed is checked by
   (*a*) test mandrel (*b*) telescope
   (*c*) wavinessmeter (*d*) spirit level
13. The test for checking straightness of the vertical movement of the knee is performed by bringing the table in central position while
   (*a*) locking table, cross-slide and the knee
   (*b*) unlocking table, cross-slide and the knee
   (*c*) locking table and cross-slide and unlocking the knee
   (*d*) unlocking table and cross-slide and locking the knee
14. In order to check and restore the accuracy of measuring instruments, ... is done.
   (*a*) alignment test (*b*) recalibration
   (*c*) grinding (*d*) restarting
15. The significance of re-calibration is:
   (*a*) It traces the measurement uncertainty of the instruments
   (*b*) It provides acceptability to the storage system
   (*c*) It reduces cost of measuring instrument
   (*d*) All of the above

CHATPER 6: LUBRICATION

1. Which of the following functions is performed by lubrication?
   (*a*) Minimization of the frictional power loss.
   (*b*) It works as coolant.
   (*c*) It also keeps machine parts clean and free from dirt.
   (*d*) All of the above
2. Example of semi-solids lubricant is
   (*a*) grease (*b*) jatropha curcas oil
   (*c*) graphite (*d*) none of these
3. Graphite, molybdenum disulfide, mica, and talc are examples of
   (*a*) liquid lubricant (*b*) solid lubricant
   (*c*) semi-solid lubricant (*d*) emulsions

4. The measure of degree of fluidity of the oil and internal resistance offered against shear of the oil film is called as
(*a*) viscosity (*b*) surface tension
(*c*) velocity (*d*) friction

5. The property of a liquid to cause bubbles and droplets due to adhesive forces among the liquid molecules at the surface of the liquid is called as.
(*a*) viscosity (*b*) surface tension
(*c*) oiliness (*d*) adhesiveness

6. The property of the lubricant by virtue of which oil can stick on to the surface of the machine parts operating under high pressure is called as
(*a*) viscosity (*b*) surface tension
c) oiliness (*d*) adhesiveness

7. The ability of the lubricant to stick to the metal surface to obstruct metal to metal contact is called as
(*a*) oiliness (*b*) viscosity
(*c*) surface tension (*d*) adhesiveness

8. A good lubricant should have the
(*a*) high viscosity (*b*) low surface tension.
(*c*) poor oiliness (*d*) poor adhesiveness.

9. A good lubricant should have the
(*a*) high flash (*b*) good oiliness and adhesiveness
(*c*) low pour point and volatility (*d*) All of these

10. The lowest temperature at which lubricating oil ceases to flow when cooled is known as
(*a*) flash point (*b*) saturation point
(*c*) dew point (*d*) pour point

11. The system of lubrication in which the load supporting film is created by the shape and relative motion of the sliding surfaces is called as
(*a*) hydrodynamic lubrication (*b*) thin film lubrication
(*c*) hydrostatic lubrication (*d*) oil to surface lubrication

12. The system of lubrication in which, the load supporting fluid, separating the two surfaces is generated by external pump is called as
(*a*) hydrodynamic lubrication (*b*) thin film lubrication
(*c*) hydrostatic lubrication (*d*) pump set lubrication

13. The condition of lubrication, in which there is partial metal-to-metal contact between sliding surfaces is called as
(*a*) boundary lubrication (*b*) thin film lubrication
(*c*) both (*a*) and (*b*) (*d*) hydrostatic lubrication

14. The type of lubrication can be seen in door hinges and machine tool slides is
(*a*) hydrodynamic lubrication (*b*) thin film lubrication
(*c*) hydrostatic lubrication (*d*) pump set lubrication

15. The type of lubricator is most useful in food or chemical processing equipments is
(*a*) screw cap lubricator (*b*) wrench type lubricator
(*c*) drop feed oil lubricator (*d*) drop cap lubricator
16. The method of lubrication used for antifriction bearings and high grade machines is
(*a*) screw cap lubricator (*b*) wrench type lubricator
(*c*) drop feed oil lubricator (*d*) drop cap lubricator
17. The type of lubricator suitable for slow speed horizontal line shafting is
(*a*) glass bottle lubricator (*b*) screw cap lubricator
(*c*) drop feed oil lubricator (*d*) ring oil lubricator
18. The type of lubricator most suited for the heavily loaded bearings is
(*a*) full pressure lubrication system (*b*) drop feed oil lubricator
(*c*) glass bottle lubricator (*d*) none of these
19. Which of the following function is performed by the engine oil?
(*a*) protect against rust and corrosion
(*b*) keep engine parts clean
(*c*) cool engine parts (*d*) All of these
20. The viscosity of multi-grade lubricating oils is improved controlled by
(*a*) viscosity-coefficient improvers (*b*) viscosity-index improvers
(*c*) both (*a*) and (*b*) (*d*) none of these

CHAPTER 7: INSPECTION AND QUALITY CONTROL

1. Who said that, "Inspection is the process of measuring the quality of a product or service in terms of established standards"
(*a*) Alfred and Beatly (*b*) Kimball
(*c*) Clausius (*d*) W.R. Speringle
2. Checking the acceptability of the manufactured product is called
(*a*) quality assurance (*b*) quality audit
(*c*) inspection (*d*) quality circle
3. In working inspection, the products should be inspected while they are
(*a*) in market (*b*) in process
(*c*) in designing stage (*d*) none of these
4. In this .... inspection, certain percentage from batch of products is inspected and the entire batch is judges from the result obtained.
(*a*) Sample (*b*) Working
(*c*) Final (*d*) Visual
5. The type of inspection in which products are checked for accuracy after completion of assembly is known as
(*a*) functional inspection (*b*) sample inspection
(*c*) design inspection (*d*) floor inspection

6. In the inspection work carried out at or near the machine in known as
(*a*) Sample inspection (*b*) Design inspection
(*c*) Endurance inspection (*d*) Floor inspection
7. Quality includes all the characteristics, which describe the
(*a*) productivity (*b*) market price
(*c*) utility (*d*) acceptability of a product
8. Planned and systematic effort by a manufacturer to assure that its products conform to the existing quality standards is known as
(*a*) quality assurance (*b*) quality circle
(*c*) quality square (*d*) quality audit
9. Small group of workers of the same work area doing similar work, who are trained to identify, analyze and solve work-related problems is known as
(*a*) quality group (*b*) quality circle
(*c*) quality triangle (*d*) quality auditors
10. The variables which are random and difficult to be identified are known as
(*a*) chance variables (*b*) assignable variables
(*c*) complex variable (*d*) none of these
11. Following are the examples of .... variable, properties of raw material, imperfection in the machine design and inherent inaccuracy of machine parts.
(*a*) assignable (*b*) complex
(*c*) chance (*d*) none of these
12. Following are the examples of …………… variables improper sequence of manufacturing operations, worn out parts of machine, defective machine parts and variation in room temperature.
(*a*) complex (*b*) assignable
(*c*) chance (*d*) none of these
13. $\overline{X}$, $\overline{R}$-charts are
(*a*) variable charts (*b*) attribute chart
(*c*) both of these (*d*) none of these
14. *p*-chart is
(*a*) variable charts (*b*) attribute chart
(*c*) both of these (*d*) none of these
15. The attribute chart is prepared for
(*a*) small industries (*b*) simple problem
(*c*) large and complex industries (*d*) all of the above
16. Types of control chart are
(*a*) complex and variable charts (*b*) variable and constant charts
(*c*) variable and attribute charts (*d*) attribute and complex charts
17. The following is not an attribute chart
(*a*) *R*-chart (*b*) *p*-chart
(*c*) *c*-chart (*d*) *u*-chart

18. Total Quality Management emphasizes
(*a*) the responsibility of the Quality Control staff to deal with all quality-related problems
(*b*) a commitment to quality that goes beyond internal company issues to suppliers and customers
(*c*) a system where strong managers are the only decision makers
(*d*) a process where mostly statisticians get involved

19. A successful TQM program incorporates all of the following except
(*a*) continuous improvement (*b*) employment involvement
(*c*) involvement of customers (*d*) centralized decision making authority

20. ISO stands for
(*a*) Indian Organization for Standardization
(*b*) International Organization for Standardization
(*c*) International Organization's Standards
(*d*) Internationally Organized Standards

CHAPTER 8: RELIABILITY AND MAINTAINABILITY

1. The reliability of the product is expressed as
(*a*) mean (*b*) ratio
(*c*) probability (*d*) none of these

2. Reliability is concerned with the performance of a product during
(*a*) manufacturing stages (*b*) lifetime
(*c*) junking time (*d*) all of the above

3. If three components are in series having reliability R_1, R_2 and R_3 respectively. Then, the overall reliability of series system is given as
(*a*) $R_1 \times R_2 \times R_3$ (*b*) $R_1 + R_2 + R_3$
(*c*) $(R_1 + R_2) \times R_3$ (*d*) $(R_1 + R_2)/R_3$

4. If two components are in parallel having reliability R_1 and R_2 respectively. Then, the overall reliability of parallel system is given as
(*a*) $R_1 \times R_2$ (*b*) $R_1 + R_2$
(*c*) $1 - (1 + R_1)(1 + R_2)$ (*d*) $1 - (1 - R_1)(1 - R_2)$

5. The ....curve shows the life cycle span of a product
(*a*) bath-tub (*b*) fault tree
(*c*) event tree (*d*) bath-tube

6. From Bath-tub curve, most of products go through .... phases from inception to junking.
(*a*) two (*b*) three
(*c*) four (*d*) five

7. .... is the probability that a product when used under stated conditions, without consideration for any maintenance, shall operate satisfactorily at any given time.
(*a*) reliability (*b*) maintainability
(*c*) quality control (*d*) availability

8. The probability that the product will be operational at any random time is called as
(*a*) reliability (*b*) maintainability
(*c*) point availability (*d*) mean availability

9. The proportion of time during a time period when the system is available for use is called as
(*a*) reliability (*b*) mean availability
(*c*) point availability (*d*) steady state availability

10. When there is a great deal of distortion of failed part, it known as
(*a*) creep (*b*) ductile failure
(*c*) brittle failure (*d*) malleable failure

11. When a part breaks with no apparent distortion, it is called as
(*a*) creep (*b*) ductile failure
(*c*) brittle failure (*d*) malleable failure

12. Fault tree, event tree and route cause analysis are the examples of
(*a*) Failure analysis methods (*b*) Root cause failure analysis
(*c*) Failure modes effect critically analysis (*d*) None of these

CHAPTER 9: MAINTENANCE

1. .... refers to application of concepts of engineering to the optimization of equipments, procedures and departmental budgets to procure better maintainability, reliability and availability of equipment.
(*a*) Reliability (*b*) Production engineering
(*c*) Maintenance engineering (*d*) Industrial management

2. This one is not the objective of maintenance
(*a*) to keep the plant, machines and equipments fit.
(*b*) to reduce the breakdown probability.
(*c*) to increase the sale of the product.
(*d*) to bring the wastages and scrap to minimum.

3. To achieve maintenance, the machines are periodically inspected for
(*a*) inadequate lubrication. (*b*) overloading of the machine tools.
(*c*) wear and tear of different parts. (*d*) all of the above

4. Development of the plans to achieve the capacity assurance objectives of an organization is the duty of
(*a*) maintenance engineer (*b*) chief engineer
(*c*) technical assistant (*d*) none of these

5. .... is defined as a process by which the management ensures that the right number and the right kind of person are selected at the right place and right time and are doing the right things for the achieving objectives of the organizational.
(*a*) Manpower planning (*b*) Entrepreneurship
(*c*) Maintenance supervision (*d*) Worker selection

6. Planning for the movement of materials and associated flow of information is defined as:
(*a*) material supervision (*b*) material procurement
(*c*) material management (*d*) material selection

7. In breakdown maintenance, the repair may be done
(*a*) randomly
(*b*) due to prediction on the basis of periodic checking
(*c*) due to the breakdown
(*d*) all of above

8. In the preventive maintenance, the repair is done
(*a*) due to prediction on the basis of periodic checking
(*b*) randomly
(*c*) due to the breakdown
(*d*) none of these

9. .... locates the cause of trouble and removes them before the breakdown occurs
(*a*) Preventive maintenance (*b*) Breakdown maintenance
(*c*) Reliability (*d*) Maintainability

10. Centralized maintenance means that those person having charge of maintenance have
(*a*) different supervisors (*b*) common supervisor
(*c*) both (*a*) and (*b*) (*d*) none of these

11. Which of the following instrument is used to measure deflection of structural parts like columns and foundation?
(*a*) Vibration meter (*b*) Shock pulse meter
(*c*) Ultrasonic flaw detector (*d*) Strain Gauge

12. .... points out defects of rolling elements like ball bearing.
(*a*) Vibration meter (*b*) Shock pulse meter
(*c*) Ultrasonic flaw detector (*d*) None of these

13. Which of the following is used to detect hot spots in bearings or other parts of the machine?
(*a*) Strain gauge (*b*) Thermo-vision
(*c*) Infrared thermometer (*d*) Resistance thermometer

14. Which instrument is used to determine the extent of choking in pipes or heat exchangers?
(*a*) Ultrasonic flow meter (*b*) Shock pulse meter
(*c*) Ultrasonic flaw detector (*d*) Pulse tube meter

15. Those techniques which determine the properties of a material without destroying them is called as
(*a*) Non-destructive testing (*b*) Non-traditional testing
(*c*) Destructive distillation (*d*) Conditioning

CHAPTER 10: MAINTENANCE PLANNING, STORAGE OF PARTS AND REPLACEMENT ANALYSIS

1. The various stages maintenance are
(*a*) inspection, servicing, repair and overhauling
(*b*) purchase, inspection and testing
(*c*) inspection, testing and replacement
(*d*) inspection, servicing and discard

2. .... is carried out to know and sense the possible defects and check the performance of various machine parts.
(*a*) Inspection (*b*) Servicing
(*c*) Repair (*d*) Overhaul

3. .... returns the machine back to its peak performance.
(*a*) Inspection (*b*) Servicing
(*c*) Repair (*d*) Overhaul

4. Tightening of nuts and bolts comes under
(*a*) inspection (*b*) repair
(*c*) servicing (*d*) overhaul

5. Complete record of machine is maintained by the maintenance department in form of
(*a*) log book (*b*) service book
(*c*) instruction manual (*d*) machine data sheet

6. The information about the repairs, modifications, replacements and overhauling conducted on the machine is maintained in
(*a*) log book (*b*) machine history card
(*c*) instruction manual (*d*) machine data sheet

7. Cannibalization of parts refer to
(*a*) resale of parts (*b*) disposal of parts
(*c*) reuse of salvageable parts (*d*) assembly of parts

8. Seiton means that all the items should place in
(*a*) alphabetic order (*b*) systematic order
(*c*) automated system (*d*) specially designed boxes

9. Which of the following practice of "5-S" practice, emphasizes on strict and continual maintenance of organization
(*a*) Seiton (*b*) Seiri
(*c*) Shitsuke (*d*) Seiketsu

10. In .... system, new items are placed in empty area, and materials are used issued from old lot.
(*a*) double area (*b*) convergent area
(*c*) gravity feed (*d*) moving division

11. In which of the following system, the new lot is place on the top of the piles and old is removed from bottom.
(*a*) double area (*b*) convergent area
(*c*) gravity feed (*d*) moving division

12. The method in which the grand total of the costs spent on buying, operating and maintaining equipment is obtained and is divided by the total estimated life to get an average annual cost is known as

(*a*) total life average method (*b*) annual cost method

(*c*) total cost method (*d*) annual cost method

CHATPER 11: ENGINEERING MATERIALS

1. The property of material by virtue of which it withstands load without fracture is called as

(*a*) Strength (*b*) Toughness

(*c*) Ductility (*d*) Brittleness

2. The property of material, which enables it to withstand shocks and vibration, is called as

(*a*) Strength (*b*) Toughness

(*c*) Hardness (*d*) Brittleness

3. The property of material by virtue of which it resists deformation under stress is known as

(*a*) Strength (*b*) Stiffness

(*c*) Hardness (*d*) Toughness

4. The property of material by virtue of which it resists abrasion and penetration is known as

(*a*) Strength (*b*) Toughness

(*c*) Hardness (*d*) Brittleness

5. The property of breaking of a material with little permanent distortion is called as

(*a*) Strength (*b*) Toughness

(*c*) Hardness (*d*) Brittleness

6. The property of material, which allows it to be beaten into thin sheets, is known as

(*a*) Strength (*b*) Malleability

(*c*) Toughness (*d*) Brittleness

7. The amount of heat per unit mass required to raise the temperature by unit degree is known as

(*a*) Specific heat (*b*) Thermal coefficient

(*c*) Thermal conductivity (*d*) Thermal expansion

8. When two dissimilar metals are joined from end to end and their junction is maintained at different temperatures, a small voltage is produced and this is known as

(*a*) Thermoelectric effect (*b*) Thermal effect

(*c*) Conduction (*d*) Radiative heat transfer

9. .... is defined as the ease with which a material can be machined.

(*a*) Machinability (*b*) Machining

(*c*) Machine coefficient (*d*) None of these

10. Pig iron is the name of

(*a*) Iron scrap (*b*) Raw material for blast furnace

(*c*) Iron containing greater percentage of carbon

(*d*) Product of blast furnace obtained by reduction of iron ore.

11. The percentage of carbon in cast iron is
(*a*) 0.25 to 0.5% (*b*) 2.5 to 5%
(*c*) 0.5% – 1.5% (*d*) 7 – 9%

12. In grey cast iron, carbon is present in the form of
(*a*) Flakes (*b*) Cementite
(*c*) Spheroids (*d*) None of these

13. The amount of carbon present in grey cast iron is
(*a*) 3 to 3.5% (*b*) 0 to 0.5%
(*c*) 0.5% – 1.5% (*d*) 7 – 9%

14. The amount of carbon present in white cast iron is
(*a*) 0 to 0.5% (*b*) 0.5% to 1.5%
(*c*) 7 to 9% (*d*) 1.75 to 2.5%

15. In nodular cast iron, the graphite is in the form of
(*a*) Nodules (*b*) Flakes
(*c*) Free carbon (*d*) Spheroid

16. Mild carbon steel contains carbon from
(*a*) 0.05 to 0.3% (*b*) 7 to 9%
(*c*) 1.5 to 2.5% (*d*) None of these

17. Medium carbon steel contains carbon from
(*a*) 0.05 to 0.3% (*b*) 7 to 9%
(*c*) 1.5 to 2.5% (*d*) 0.3 to 0.6%

18. High carbon steel contains carbon from
(*a*) 0.05 to 0.3% (*b*) 7 to 9%
(*c*) 0.6 to 1.5% (*d*) 0.3 to 0.6%.

19. 18 – 4 – 1 HSS contains
(*a*) 18% tungsten, 4% chromium and 1% vanadium
(*b*) 18% chromium, 4% tungsten and 1% vanadium
(*c*) 18% vanadium, 4% chromium and 1% tungsten
(*d*) None of these

20. Stainless steel is so called because of its
(*a*) high strength (*b*) high corrosion resistance
(*c*) high ductility (*d*) brittleness

21. The purpose of heat treatment is to change
(*a*) micro-structure of the materials and improves its mechanical properties.
(*b*) chemical composition
(*c*) corrosion properties
(*d*) surface finish

22. Case hardening is the method suitable for hardening
(*a*) high carbon steel (*b*) mild steel
(*c*) low carbon steel (*d*) high alloy steel

23. In nitriding steel parts, the following atmosphere is used in the furnace
(*a*) inert (*b*) nascent nitrogen
(*c*) ammonia (*d*) carbon

24. A small portion of the workpiece can be hardened by
(*a*) nitriding (*b*) case hardening
(*c*) cyaniding (*d*) flame and induction hardening

25. The purpose of tempering is to relieve the
(*a*) stress and brittleness (*b*) corrosion resistance
(*c*) strength (*d*) surface finish

26. Brass is an alloy of
(*a*) copper and iron (*b*) copper and zinc
(*c*) copper and Aluminium (*d*) copper and tin

27. Cartridge brass contains
(*a*) 28 to 38% zinc (*b*) 4 to 10% zinc
(*c*) 0.5 to 15% zinc (*d*) 50% zinc

28. Muntz metal contains
(*a*) 10% zinc (*b*) 40% zinc
(*c*) 20% zinc (*d*) 80% zinc

29. Bronze is an alloy of
(*a*) silver (*b*) tin
(*c*) copper (*d*) aluminium

30. The alloy of aluminium containing 94% aluminium, 4% copper, 0.5% manganese, 0.5% magnesium, 0.5% iron and 0.5% silicon is
(*a*) duralumin (*b*) *Y*-alloy
(*c*) magnalium (*d*) bauxite

31. Y-alloys is the alloy of
(*a*) aluminium and copper (*b*) copper and silver
(*c*) aluminium and silver (*d*) iron and aluminium

32. Magnalium contains
(*a*) aluminium, magnesium and silver (*b*) aluminium, silver and copper
(*c*) aluminium, magnesium and copper (*d*) None of these

33. Hindalium is an alloy of
(*a*) aluminium and copper (*b*) aluminium and magnesium
(*c*) aluminium and silver (*d*) copper and magnesium

34. Melting point of nickel is
(*a*) 500 ºC (*b*) 250 ºC
(*c*) 1450ºC (*d*) 2500 ºC

35. When 67% nickel is alloyed with 30% copper and 3% aluminium, it produces an alloy called as
(*a*) muntz metal (*b*) babbit metal
(*c*) brass (*d*) monel metal

36. Nichrome contains
(*a*) nickel, chromium and iron (*b*) nickel, silver and iron
(*c*) nickel, copper and zinc (*d*) nickel, zinc and iron

37. The heating elements of electric irons are made of
(a) copper (b) nichrome
(c) constantan (d) aluminium

38. The plastics which can be softened by heating and moulded into shape again are known as
(a) thermoplastics (b) thermosetting plastics
(c) epoxy (d) none of these

39. Polyethylene, polystyrene and polyvinyl chloride are the examples of:
(a) thermosetting plastics (b) rubber
(c) thermoplastics (d) none of these

40. Those plastics which cannot be softened and reworked are known as
(a) thermosetting plastics (b) thermoplastics
(c) rubber (d) none of these

CHAPTER 12: CORROSION-ITS CONTROL AND CHEMICAL CLEANING

1. The deterioration of materials by chemical interaction with their environment is known as
(a) priming (b) corrosion
(c) chemical affinity (d) none of these

2. The rate of corrosion is expressed as the loss in weight of specimen
(a) per unit surface area per unit time (b) per unit volume per unit time
(c) per unit volume (d) per unit time

3. Corrosion of metals involves
(a) physical reactions (b) chemical reactions
(c) both (d) none of these

4. The following factors play vital role in corrosion process
(a) temperature (b) climate
(c) both (d) none of these

5. Dry corrosion involves
(a) reaction with gases at elevated temperature
(b) reaction with atmosphere
(c) reaction with inert gases (d) none of these

6. Wet corrosion takes place
(a) due to high temperature
(b) in presence of water or a conducting liquid
(c) in presence of hot gases (d) all of the above

7. The speed of electrochemical attack is .... in a hot climate.
(a) increased (b) decreased
(c) remains same (d) none of these

8. The type of corrosion, which occurs as the result of a combination of an aggressive chemical environment and high fluid-surface velocities is called as
(a) inter-granular corrosion (b) erosion corrosion
(c) galvanic corrosion (d) stress concentration

9. Which of the following can be used for galvanizing
   (*a*) cobalt (*b*) copper
   (*c*) zinc (*d*) all the above
10. The method to prevent corrosion of iron by zinc coating is called
   (*a*) galvanization (*b*) electrolysis
   (*c*) cathode protection (*d*) anode protection
11. Pitting corrosion is due to
   (*a*) breaking of the protective layer of the metal
   (*b*) acidity
   (*c*) stress in the metal (*d*) All the above
12. For the cleaning of organic oils, which of the following should be used
   (*a*) di-cholorethylene (*b*) naptha and tri-choloroethylene
   (*c*) both of these (*d*) none of these
13. The colored glassy substance, transparent or opaque, fused to the surface of object is known as
   (*a*) enamel (*b*) galvanized iron
   (*c*) zinc paint (*d*) snamel
14. When machine parts are to be transported, a layer of grease is applied to them. This method is called as
   (*a*) slushing (*b*) enamelling
   (*c*) etching (*d*) none of these
15. A silvery-white, frosted appearance can be given to aluminium by applying a caustic solution to its surface in a process called as
   (*a*) chemical etching (*b*) slushing
   (*c*) galvanization (*d*) enamelling
16. Machine parts are generally protected by
   (*a*) curved shape (*b*) polished surfaces
   (*c*) different metal parts having nearly the same electrode potential are joined
   (*d*) all are applicable

KEY TO OBJECTIVE QUESTIONS

Chapter 1: Safety Practices

1. (*c*)	2. (*b*)	3. (*d*)	4. (*d*)	5. (*a*)
6. (*b*)	7. (*d*)	8. (*a*)	9. (*d*)	10. (*a*)

Chapter 2: System of Limits and Fits

1. (*a*)	2. (*c*)	3. (*a*)	4. (*d*)	5. (*c*)
6. (*b*)	7. (*c*)	8. (*d*)	9. (*b*)	10. (*a*)
11. (*b*)	12. (*c*)	13. (*a*)	14. (*b*)	15. (*c*)
16. (*d*)	17. (*b*)	18. (*a*)	19. (*b*)	20. (*a*)

Chapter 3: Installation of Machine Tools

1. (*a*)	**2.** (*b*)	**3.** (*c*)	**4.** (*d*)	**5.** (*a*)
6. (*c*)	**7.** (*b*)	**8.** (*c*)	**9.** (*a*)	**10.** (*c*)
11. (*b*)	**12.** (*a*)	**13.** (*c*)	**14.** (*a*)	**15.** (*b*)
16. (*a*)	**17.** (*a*)	**18.** (*a*)	**19.** (*b*)	**20.** (*c*)

Chapter 4: Miscellanous Examples of Installation

1. (*d*)	**2.** (*a*)	**3.** (*c*)	**4.** (*b*)	**5.** (*c*)
6. (*a*)	**7.** (*d*)	**8.** (*c*)	**9.** (*b*)	**10.** (*c*)
11. (*b*)	**12.** (*a*)			

Chapter 5: Machine Tool Testing

1. (*c*)	**2.** (*b*)	**3.** (*a*)	**4.** (*c*)	**5.** (*a*)
6. (*d*)	**7.** (*a*)	**8.** (*c*)	**9.** (*b*)	**10.** (*c*)
11. (*a*)	**12.** (*d*)	**13.** (*c*)	**14.** (*b*)	**15.** (*a*)

Chapter 6: Lubrication

1. (*d*)	**2.** (*a*)	**3.** (*b*)	**4.** (*a*)	**5.** (*b*)
6. (*c*)	**7.** (*d*)	**8.** (*a*)	**9.** (*d*)	**10.** (*d*)
11. (*a*)	**12.** (*c*)	**13.** (*c*)	**14.** (*b*)	**15.** (*a*)
16. (*c*)	**17.** (*d*)	**18.** (*a*)	**19.** (*d*)	**20.** (*b*)

Chapter 7: Inspection and Quality Control

1. (*d*)	**2.** (*c*)	**3.** (*b*)	**4.** (*a*)	**5.** (*a*)
6. (*d*)	**7.** (*d*)	**8.** (*a*)	**9.** (*b*)	**10.** (*a*)
11. (*c*)	**12.** (*b*)	**13.** (*a*)	**14.** (*b*)	**15.** (*c*)
16. (*c*)	**17.** (*a*)	18. (*b*)	**19.** (*d*)	**20.** (*b*)

Chapter 8: Reliability and Maintainability

1. (*c*)	**2.** (*b*)	**3.** (*a*)	**4.** (*d*)	**5.** (*a*)
6. (*b*)	**7.** (*d*)	**8.** (*c*)	**9.** (*b*)	**10.** (*b*)
11. (*c*)	**12.** (*a*)			

Chapter 9: Maintenance

1. (*c*)	**2.** (*c*)	**3.** (*d*)	**4.** (*a*)	**5.** (*a*)
6. (*c*)	**7.** (*c*)	**8.** (*a*)	**9.** (*a*)	**10.** (*b*)
11. (*d*)	**12.** (*b*)	**13.** (*c*)	**14.** (*a*)	**15.** (*a*)

Chapter 10: Maintenance Planning, Storage of Parts and Replacement Analysis

1. (*a*)	**2.** (*a*)	**3.** (*b*)	**4.** (*a*)	**5.** (*d*)
6. (*b*)	**7.** (*c*)	**8.** (*b*)	**9.** (*d*)	**10.** (*a*)
11. (*c*)	**12.** (*a*)			

Chapter 11: Engineering Materials

1. (*a*)	2. (*b*)	3. (*b*)	4. (*c*)	5. (*d*)
6. (*b*)	7. (*a*)	8. (*a*)	9. (*a*)	10. (*d*)
11. (*b*)	12. (*a*)	13. (*a*)	14. (*d*)	15. (*a*)
16. (*a*)	17. (*d*)	18. (*c*)	19. (*a*)	20. (*b*)
21. (*a*)	22. (*c*)	23. (*c*)	24. (*d*)	25. (*a*)
26. (*b*)	27. (*a*)	28. (*b*)	29. (*c*)	30. (*a*)
31. (*a*)	32. (*c*)	33. (*b*)	34. (*c*)	35. (*d*)
36. (*a*)	37. (*b*)	38. (*a*)	39. (*c*)	40. (*a*)

Chapter 12: Corrosion–Its Control and Chemical Cleaning

1. (*b*)	2. (*a*)	3. (*b*)	4. (*c*)	5. (*a*)
6. (*b*)	7. (*a*)	8. (*b*)	9. (*c*)	10. (*a*)
11. (*a*)	12. (*b*)	13. (*a*)	14. (*a*)	15. (*a*)
16. (*d*)				

❑❑❑

Experiments

EXPERIMENT NO. 1

OBJECT

Preparation of Preventive Maintenance Check

INTRODUCTION

Preventive Maintenance is interchangeably termed as planned maintenance, scheduled maintenance or systematic plant maintenance. This is an important function as it reduces maintenance cost, keeps the equipment in good operational condition and thus, increases its reliability.

Preventive maintenance locates the cause of trouble and removes them before the breakdown occurs. Thus, it justifies the idea that prevention is better than cure. Imperative safeguard against costly breakdown is to inspect, lubricate and check the equipment before time as frequently as possible. In order, to utilize the equipment to the full in reliable condition, necessary measures may be taken to maintain it and to prevent overloading, dampness, negligence and its misuse. Frequent inspection may be done on the basis of its usefulness, importance and delicacy. Thus, periodic inspections or assessment modes help to find out the reasons, leading to breakdown and to rectify them, when they are in initial stages. This type of repair needs less time as compared to that of breakdown repair. The preventive maintenance actually saves time and increases production. Preventive maintenance can be carried out weekly, monthly or annually. Card or Chart is used to maintain records of preventive maintenance check.

CLASSIFICATIONS OF PREVENTIVE MAINTENANCE CARDS

There are three types of cards for preventive maintenance:

(*i*) Weekly Preventive Maintenance Card: It includes:

- (*a*) Cleaning of machine/equipment
- (*b*) Check for lubricating oil
- (*c*) Check for coolant
- (*d*) Filter check
- (*e*) Minor adjustments (like tightening of nuts/bolts etc)

(*ii*) Monthly Preventive Maintenance Card: It includes the works done in weekly maintenance and following additional work:

- (*a*) Belt change
- (*b*) Check for all axis movement and alignment

(*iii*) Annual Preventive Maintenance Card: It includes the works done in monthly maintenance and following additional work:

(*a*) Machine level check

(*b*) Replacement of lubricating oil

The following steps are to be followed by the operator while performing preventive maintenance on the lathe

(*a*) Cleaning of machine

(*b*) Clean the chips from the machine surfaces

(*c*) Check the level of the cutting fluid reservoir

(*d*) Lubricate the bed ways daily with light oil

(*e*) Check the oil level in headstock periodically

(*f*) Check the apron oil level regularly

The various cards of Preventive maintenance for lathe machine are shown below.

Weekly Maintenance Card					
Name of the Department:					
Name of the Machine:		Model/Make:		Year:	
For the .... week of .... month of .... year					
Name of the operator:					
Job description	Lead Screw	Head Stock	Saddle	Speed box	Motor bearing
1. Lubrication					
2. Check for coolant					
3. Filter check					
4. Minor adjustments					
Checked by					

Monthly Maintenance Card		
Name of the Department:		
Name of the Machine:	Model/Make:	Year:
For the .... week of .... month of .... year		
Name of the operator:		
Job Description	Operation Performed	
1. Belt Change		
2. Check for coolant		
3. Check for movement of axis		
4. Check for alignment of axis		
Checked by		

Annual Maintenance Card		
Name of the Department:		
Name of the Machine:	Model/Make:	Year:
For the .... week of .... month of .... year		
Name of the operator:		
Job Description	Operation Performed	
1. Spindle axis alignment 2. Machine level check 3. Spindle bearing adjustment 4. Replacement of lubricating oil		
Checked by		

EXPERIMENT NO. 2

OBJECT

Condition Monitoring by NDT

THEORY

Condition Monitoring (CM) is the process of determining condition of machines or equipment, such as any change is indicative of development failure. It is an important component of predictive maintenance technique. The application of condition monitoring allows maintenance to be planned so that required action may be taken in order to avoid failure.

Machines with defects are more likely to fail than defect free machines. Once a defect has been identified, the failure development has already started and condition monitoring can only measure the deterioration of the condition. Intervention in the initial phases of deterioration is more cost effective than allowing the machine to fail completely. In today's competitive world, significant savings can be achieved by application of CM.

Non-destructive testing (NDT) includes wide group of analysis techniques used in industry to determine the properties of a material, or system without causing any destruction. NDT technique saves money as well as time. Also, it does not change the properties of the system under testing. Common NDT methods include ultrasonic, magnetic-particle, liquid radiographic, visual inspection eddy-current testing etc. It finds wide applications in field of engineering and medicine.

PROCEDURE

Based on NDT, some important types of condition monitoring techniques are:

(*i*) Infrared thermography

(*ii*) Alignment of shafts using LASER

(*iii*) Oil analysis

(*iv*) Leakage detection

(*v*) Ultrasonic crack detection

(*vi*) Ultrasonic thickness monitoring

Infrared Thermography: Infrared thermography is also called as thermal imaging. It detects infrared radiations emitted from object, converts it to temperature and displays image of temperature distribution. Temperature rise can be sources of number of problems in any machine or equipment. Condition monitoring by using this technique gives a quick and safe way of detecting problems in such situations. At the same time, it can be carried out while machine is under operation. Temperature measurement can be done from a distance without coming in contact with the machine.

Alignment of Shafts using LASER: Perfect alignment of the shafts is very important for high speed rotating machine since misalignment can cause failure of machine. LASER shaft alignment system utilizes LASER-based technology for checking alignment of shafts. LASER shaft alignment systems with wireless Bluetooth operation, color graphics and infinite battery life are available. Such system offers accurate and precision alignment of shafts within hundredths of mm.

Oil Analysis: Contamination in oil can be detected by means of condition monitoring by **oil analysis.** There can be different reasons of contamination such as presence of water particles, carbon deposits, traces of other oil etc. One of the techniques of condition monitoring through oil analysis is oil circuit and transformers where condition of the oil is monitored for the presence of carbon deposits.

Leakage Detection: Leakage is unintentional discharge of gases from mechanical equipments such as pressure vessels, pipelines, pressure containment parts, aerosol containers etc. Detection of colorless and odorless gases is a tedious task. The purpose of leak detection systems is to detect and locate leakage. Such system provides an alarm and displays other important information. Leaks can be detected by using liquid/gas penetrant techniques, electronic listening devices or bubble tests etc.

Ultrasonic Crack Detection: Detection of minute cracks is a difficult task specially if the cracks are near other inhomogeneities, pores, material inclusions or geometric features. The laws of physics that govern the propagation of sound waves through solids is used to detect minute cracks. High frequency sound waves are reflected from crack in predictable ways, producing distinctive sound patterns which is displayed and recorded. Ultrasonic crack detection technique finds application in manufacturing industries involving welds and structural metals.

Ultrasonic Thickness Monitoring: Ultrasonic thickness monitoring is quite an effective technique adopted for finding out wall thickness of corroded metals like steam pipes, water pipes, storage tanks, pressure vessels etc. It is also used for determination of thickness of metals, plastics, ceramics and glass etc. Measurements can be made from single side only. Ultrasonic thickness gauges work by measuring how long it takes for a sound pulse to travel through material under test. The ultrasound waves inside a material will bounce back from the opposite surface like an echo. Thus, thickness measurements can be made instantly and accurately. Ultrasonic thickness measurements save material, time and costs.

Advantages of condition monitoring

The advantages of condition monitoring are as follows:

1. Maintenance can be planned
2. It maximizes availability and reliability of machine
3. It minimizes unscheduled downtime
4. Cause of failure can be analyzed
5. It improves operating efficiency
6. It reduces maintenance costs (better planning)
7. It improves customer relations

Disadvantages of condition monitoring

The disadvantages of condition monitoring are as follows:

1. It requires careful choice of the correct technique
2. Monitoring machine cost is high
3. Skilled personnel needed
4. Strong management commitment needed

EXPERIMENT NO. 3

OBJECT

Case Study of Trouble Free Maintenance

CASE STUDY OF HYDRUALIC PUMP

Teacher may ask the students to conduct a case study of trouble free maintenance of any machine or equipment and subsequently to write report on it. The experimental sheet presents such a case study of trouble free maintenance of hydraulic pump.

TROUBLE FREE MAINTENANCE OF PUMPS

A maintenance engineer adopts various techniques for trouble free operation of pumps and thereby increases availability of the pump. These techniques include routine maintenance, preventive and breakdown maintenance, periodic overhaul and condition monitoring. If the maintenance personnel are busy doing troubleshooting maintenance and repair work, it is not very praiseworthy.

Some pumps get branded as troublesome right from the beginning. The twin reasons may be improper selection or improper installation of the pump. The other reason for pump trouble is improper operation. Thus, initially trouble-free maintenance of pumps depends upon various stages of selection, installation and operation of pump. Basically, the entire technologies of pump construction, its designing, protection techniques used, instruments used and correct installation method should be correctly explored to get the optimum performance from the pump with least the maintenance.

If the pump is shut-down for repair work or overhaul, the down time increases. Inventory of replacement of parts is also an associated point of concern. In the event of operational failure of the pump, diagnosis of the cause of failure should be investigated using proper failure analysis. Thus, it is clear that the prevention of frequent recurrence of such failure can be assured in the process of repair itself.

The main mantra for a trouble-free operation of pump should be "keep it like a new pump". It is necessary to give special attention focusing on cleanliness, ventilation and cooling. By keep pump clean, leakage can also be easily detected. All the pipe joints should be regularly inspected. Fitting of impeller and shaft is to be checked. Proper cooling is necessary to avoid abnormal temperature build-up. Pump should be regularly lubricated as per manufacture's guideline.

Mechanical vibrations and abnormal noise should also be continuously monitored. Pump misalignment or loose bolts can cause mechanical vibrations. Bolts should be tightened. Common leakage problem of pump is at shaft seal which may take place due to misalignment. Therefore, the alignment of pump and shaft should be checked regularly.

It is also imperative to check the functioning of all valves. As long as the bearings are good, the rotation of the pump will be in proper order. Thus, the life of the pump depends on conditions of bearings. Worn out bearings should be replaced immediately. The casing of the pump should also be taken care of, if necessary. The paint of the pump is not just for beautification. It acts like protective coating, especially in case of corrosion friendly environments. Maintenance worker should also be familiar with the colors of wires in a three-phase power supply. Therefore, connections should be properly done.

The manufacture's instruction manual should be read properly for installing and operating pump. These manuals also contain information of common pump troubles. In case of problem, maintenance engineer should first check the cause mentioned in the instruction manuals.

EXPERIMENT NO. 4

OBJECT

Project on maintenance of utility equipment like compressors and pumps

MAINTENANCE STAGES

The various stages of cumulative maintenance jobs are known as maintenance stages. These stages are Inspection, Servicing, Repair and Overhauling (ISRO).

MAINTENANCE STAGES OF RECIPROCATING COMPRESSORS

An air compressor is a machine which compresses the air and increases its pressure. It inducts air from the atmosphere, compresses it and then delivers it to a storage vessel. From where, it is conveyed through the pipeline to a place where the supply of compressed air is required. The compression of air requires work to be done on it. Thus, compressor is always equipped with a prime mover like an electric motor. The power available from compressed air is used in many applications.

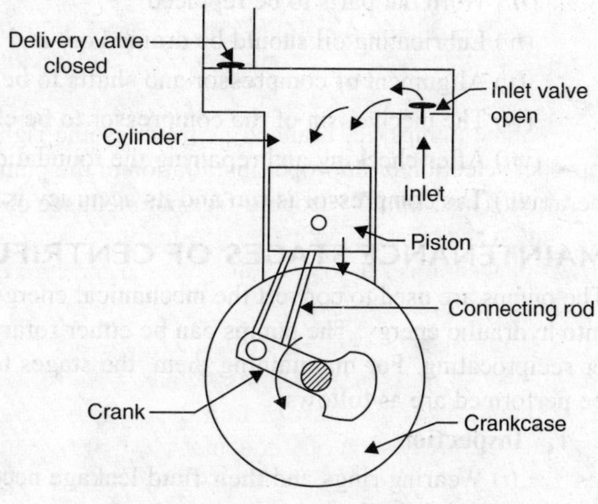

Figure E-4.1 Reciprocating Air Compressor

For maintaining them, the stages to be performed are as follows:

1. **Inspection**
   (*i*) Oil leakages need to be checked.
   (*ii*) Oil level in the crankcase need to be checked.
   (*iii*) Tightening of the bolts and replacing broken/worn bolts.
   (*iv*) All joints to be carefully checked for leakage.
   (*v*) Lubricating oil pressure must be checked.
   (*vi*) The alignment of compressor and shaft is checked.
2. **Servicing**
   (*i*) Some of the parts are disassembled.
   (*ii*) These parts are washed/cleaned and lubricated.
   (*iii*) Surface of crank pin and journal should be checked.

(*iv*) Clearance between the crank pin and main bearing should be checked and compared with prescribed data.

(*v*) Cylinder, piston and piston rod need to be checked.

(*vi*) Functioning of valves should be checked.

(*vii*) For next stage, a list of defects is to be prepared.

3. **Repair**

(*i*) Major units and sub-units should be dissembled. They are washed and cleaned.

(*ii*) Worn out bearings should be replaced.

(*iii*) All moving parts should be checked for wear and replaced, if required.

(*iv*) All the nuts and bolts should be tightened. Worn out nuts and bolts should be replaced.

(*v*) Functioning of valves should be checked.

(*vi*) Lubricating oil should be drained out and fresh oil should be re-filled.

(*vii*) The relative position of parts in sub-assemblies be checked and then fixed.

(*viii*) Paint sub-units, if required before re-assembly.

4. **Overhaul**

(*i*) The compressor should be disassembled. All the parts should be washed and cleaned.

(*ii*) All the nuts and bolts should be tightened. Worn out nuts and bolts should be replaced.

(*iii*) Worn out parts to be replaced.

(*iv*) Lubricating oil should be drained out and fresh oil should be re-filled.

(*v*) Alignment of compressor and shafts to be checked.

(*vi*) The mechanism of the compressor to be checked for its relative motion.

(*vii*) After checking and repairing the foundation, the parts are painted and reassembled.

(*viii*) The compressor is run and its accuracy is checked.

MAINTENANCE STAGES OF CENTRIFUGAL PUMPS

The pumps are used to convert the mechanical energy into hydraulic energy. The pumps can be either rotary or reciprocating. For maintaining them, the stages to be performed are as follows:

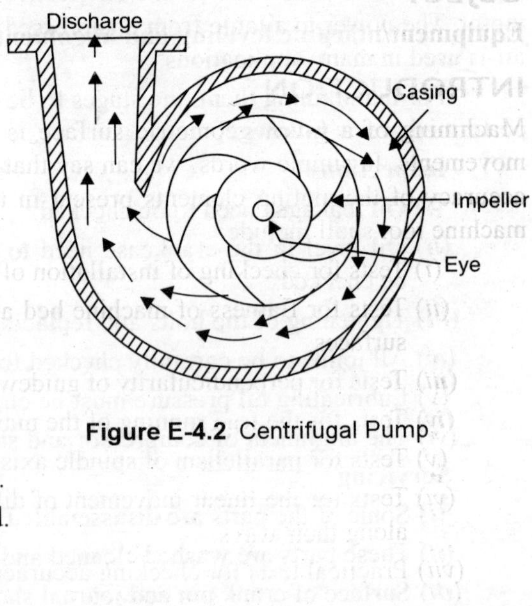

Figure E-4.2 Centrifugal Pump

1. **Inspection**

(*i*) Wearing rings and their fluid leakage need to be checked.

(*ii*) Tightening of the bolts and replacing broken/ worn bolts.

(*iii*) Stuffing box and leakage of fluid from them be checked.

(*iv*) All the pipe joints to be checked.

(*v*) The lubrication of bearings is inspected.

(*vi*) The alignment of pump and shaft is checked.

(*vii*) Fitting of impeller and shaft is to be checked.

2. **Servicing**

(*i*) Some of the parts are disassembled.

(*ii*) They are washed/cleaned and lubricated.

(*iii*) Worn out lining or bushes in bearing are replaced.

(*iv*) The strainer is to be washed/cleaned.

(*v*) Functioning of all valves should be checked.

(*vi*) For next stage, a list of defects is to be prepared.

3. **Repair**

(*i*) Major units and sub units linkages be disassembled.

(*ii*) Worn out bearings should be replaced.

(*iii*) The casing of the pump should be repaired, if required.

(*iv*) All the nuts and bolts should be tightened. Worn out nuts and bolts should be replaced.

(*v*) The coupling bushes, if necessary should be repaired/replaced.

(*vi*) The relative position of parts in sub-assemblies be checked and fix them according to the drawings to give excellent packing.

(*vii*) The worn out packing of stuffing box should be replaced.

4. **Overhaul**

(*i*) The pump should be disassembled.

(*ii*) All the parts should be washed and cleaned.

(*iii*) The bearings and packing of pumps should be changed.

(*iv*) Alignment of shafts is checked.

(*v*) The worn out pipe fitting is replaced.

(*vi*) After checking and repairing the foundation, the parts are painted and reassembled.

(*vii*) The pump is run and its accuracy is checked.

EXPERIMENT NO. 5

OBJECT

Equipment/machine leveling and alignment

INTRODUCTION

Machining of a given geometric surface is achieved primarily by the combination of work-tool movements. In simple words, we can say that the accuracy of the machined surface depends upon the accuracy of the mating elements present in the machine tool. The different tests conducted on any machine tool shall include:

(*i*) Tests for checking of installation of machine in horizontal or vertical directions.

(*ii*) Tests for flatness of machine bed and straightness and parallelism of bed ways or bearing surfaces.

(*iii*) Tests for perpendicularity of guideways with respect to other guideways or bearing surfaces.

(*iv*) Tests for the true running of the main spindle and its axial movements.

(*v*) Tests for parallelism of spindle axis with respect to guideways or bearing surfaces.

(*vi*) Tests for the linear movement of different members e.g., saddle and table cross-slides etc., along their ways.

(*vii*) Practical tests for checking accuracy and surface finish of the product being machined.

Dial indicator, tests mandrels, spirit levels, straight edges and squares, autocollimators, alignment telescope, water level, wavinessmeter etc. are the instruments needed to conduct these tests.

ALIGNMENT TETS ON LATHE

The some of the important alignment tests performed on the lathe machine are as follows:

1. Level of the Machine: Before conducting test on any machine tool, it is necessary that it should be installed correctly in horizontal and vertical planes on properly constructed foundation. It is essential that it must be isolated from the surrounding by using vibration dampers.

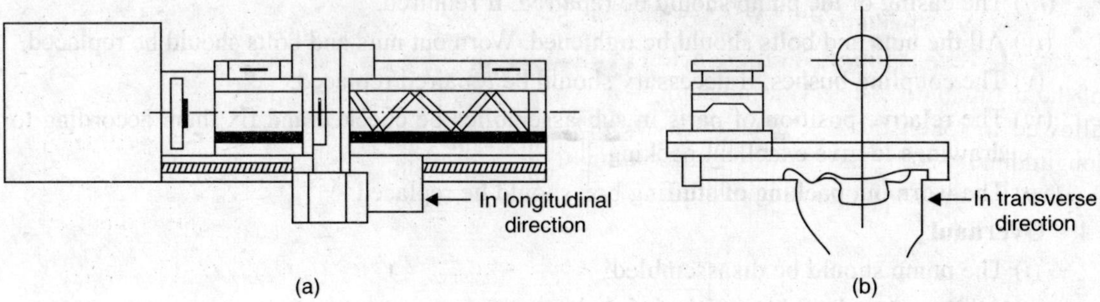

Figure E-5.1 Level of installation

The level of the machine bed in longitudinal and transverse directions is checked by a spirit level. The spirit level is placed over the gauge block on the bedways at the front position, back position and in the cross wise direction. The position of the bubble in the spirit level is checked and the readings are noted. The setup for testing the level of a lathe machine is shown in Figure E-5.1. Permissible error for front guideways is 0.02 mm/m convex while for rear guideways, it is 0.01 to 0.02 mm/m convex.

2. Straightness of saddle in horizontal plane: The test mandrel is held between two centres. The dial indicator is mounted on the saddle. The plunger of the dial indicator is made to touch the test mandrel as shown in Figure E-5.2.

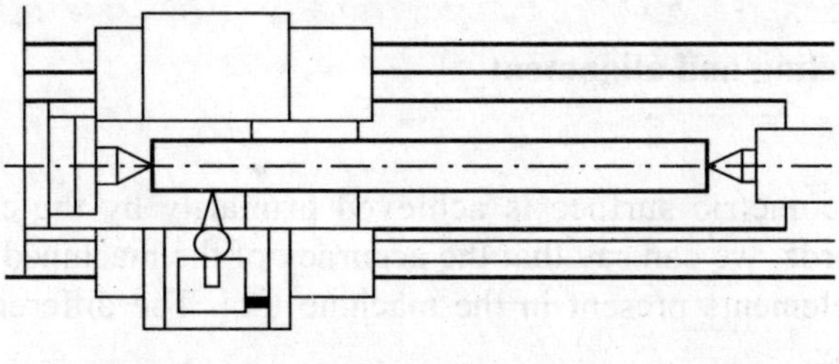

Figure E-5.2 Straightness of the saddle

The saddle is then moved longitudinally along the length of the mandrel. Readings are taken at various places. Permissible error is 0.02 mm over the length of the mandrel.

3. Alignment of both centres in the vertical plane: This test is carried out in vertical place. The test mandrel is fitted between two centres. The dial indicator is mounted on the saddle in vertical place. The plunger of the dial indicator is allowed to touch the test mandrel as shown in Figure E-5.3. The saddle along with the dial indicator is travelled longitudinally along the bedways over the length of the mandrel. The readings are taken at various places. Permissible error is 0.02 mm over the length of the mandrel.

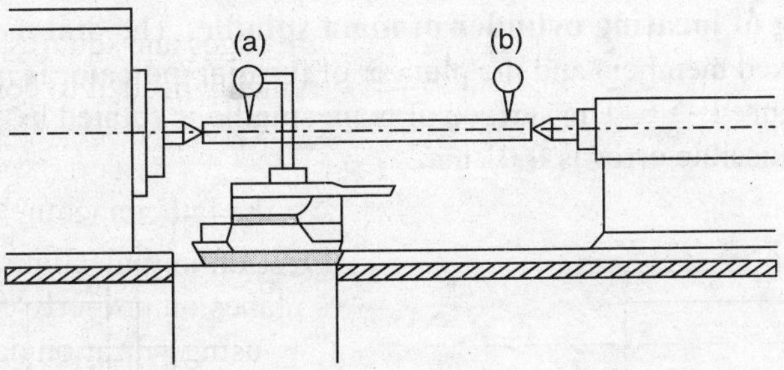

Figure E-5.3 Alignment of both centers in the vertical plane

4. True running of taper socket in main spindle: The test mandrel is fitted into tapered socket of the main spindle. The dial indicator is mounted in the saddle. The plunger of the dial indicator is allowed to touch the test mandrel as depicted in the Figure E-5.4. The saddle is then travelled longitudinally along the bedways and readings are taken at the point (*a*) and (*b*). Permissible error at position (*a*) is 0.01 mm and at position (*b*), it is 0.02 mm.

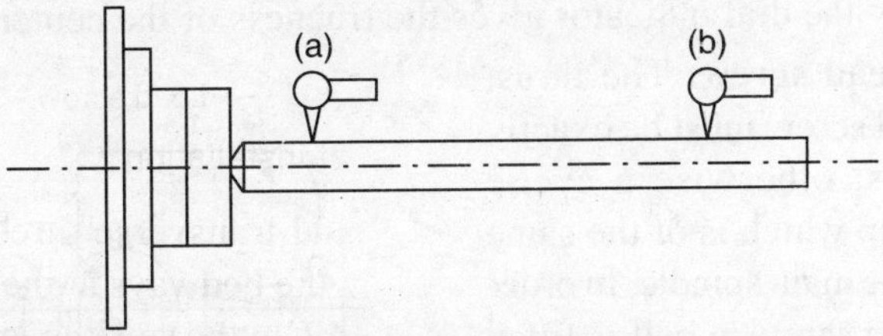

Figure E-5.4 True running of taper socket in main spindle

5. Parallelism of main spindle to saddle movement

The test mandrel is fitted in the tapered socket of the spindle. The dial indicator is mounted on the saddle. The plunger of the dial indicator is allowed to touch the test mandrel and the saddle is moved to and fro. It is checked in vertical as well as in horizontal plane. The indication in horizontal plane is given by (*b*) and in vertical plane by (*a*). Permissible error is 0.02 mm per length of test mandrel rising towards free end only and 0.02 mm per length of test mandrel inclined at free end towards tool pressure only.

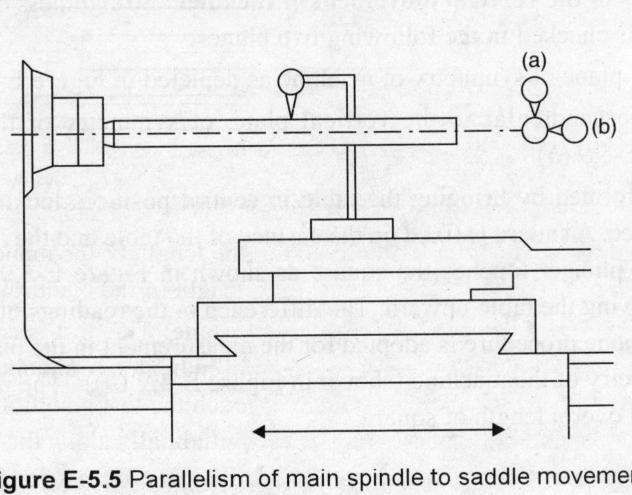

Figure E-5.5 Parallelism of main spindle to saddle movement

6. True running of locating cylinder of main spindle: The dial indicator is mounted on the carriage (or any other fixed member) and the plunger of the dial indicator is made to touch the locating cylinder as shown in Figure E-5.6. The surface of main spindle is rotated by hand and readings of dial indicator are taken. Permissible error is 0.01 mm.

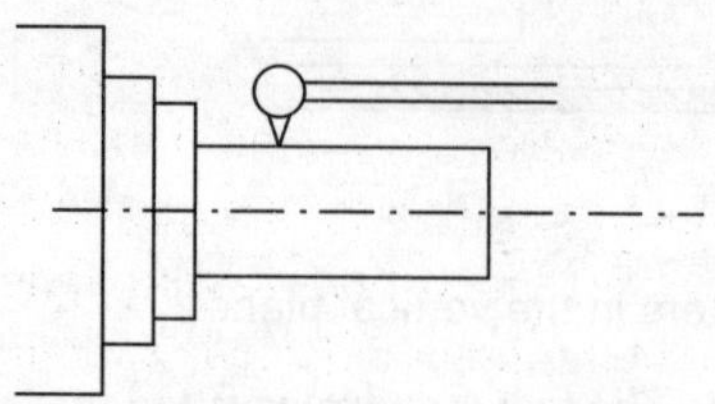

Figure E-5.6 True running of locating cylinder of main spindle

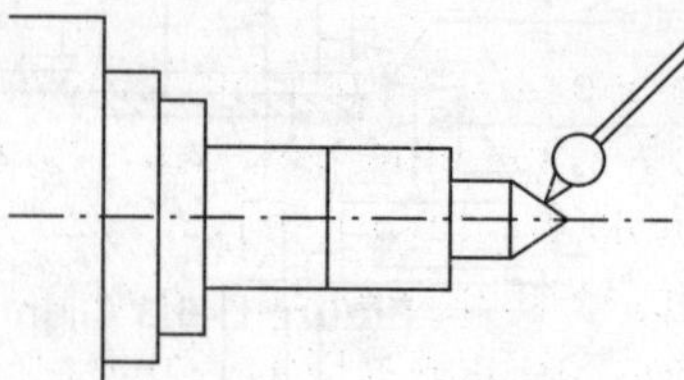

Figure E-5.7 True running of head stock center

7. True running of head stock center: The plunger of the dial indicator is made to touch perpendicular to the tapered surface of the center as shown in Figure E-5.7 and the spindle is rotated. The deviation indicated by the dial indicator gives the trueness of the center.

8. Axial slip of lead screw: The thrust face and collars of the lead screw must be exactly square to the screw axis, otherwise a cyclic endwise movement is setup which is of the same nature as the axial slip in the main spindle. In order to test axial slip in the lead screw, a ball is fitted in the end of lead screw, and the plunger of the dial indicator is pressed against the ball as shown in Figure E-5.8. The lead screw is rotated and deviation, if any, is noted.

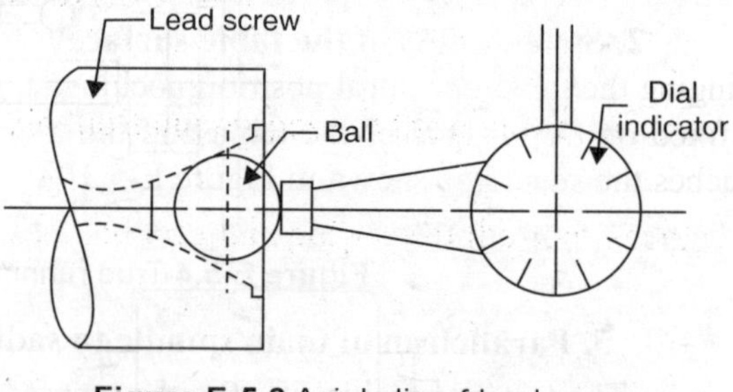

Figure E-5.8 Axial slip of lead screw

ALIGNMENT TEST ON MILLING MACHINE

Some of the important alignment tests performed on milling machine are as follows:

1. **Straightness of the vertical movement of the knee:** Straightness of the vertical movement of the knee is checked in the following two planes:

   (*a*) Vertical plane of symmetry of machine as depicted in Figure E-5.9 (*a*).

   (*b*) Plane perpendicular to the vertical plane of symmetry of the machine depicted in Figure E-5.9 (*b*).

This test is performed by bringing the table in central position, locking table and cross-slide while unlocking the knee. A square is fixed on the surface of the table and the dial indicator is mounted in such a way that its plunger touches the square as shown in Figure E-5.9 (*a*). The dial indicator reading is noted by moving the table upward. The difference in the readings of the dial indicator is the measure of error. The same procedure is adopted for the measurement in the plane perpendicular to the vertical plane of symmetry of the machine (shown in Figure E-5.9 (*b*)). The permissible error in both the planes is 0.025 mm over a length of square.

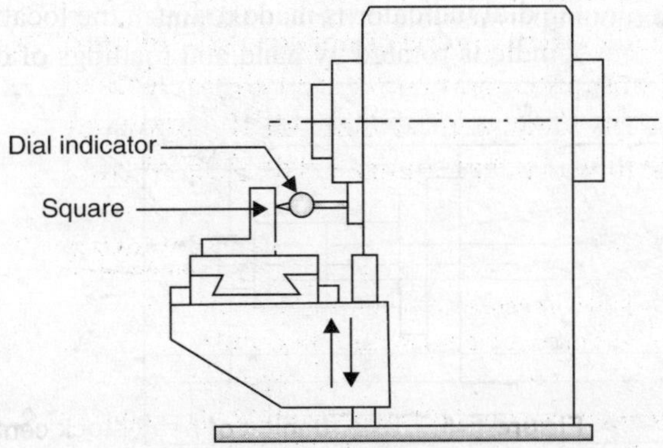

Figure E-5.9 (*a*) Straightness of the vertical movement of the knee in the vertical plane of symmetry.

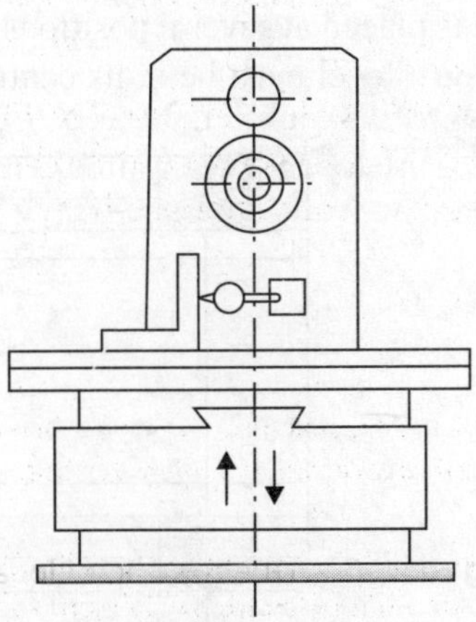

Figure E-5.9 (*b*) Straightness of the vertical movement of the knee in the plane perpendicular to the vertical plane of symmetry

2. Straightness of the table surface to the column ways for knee: This test is performed by bringing the table in central position, locking table and cross-slide while unlocking the knee. A square is fixed on the surface of the table and the dial indicator is mounted in such a way that its plunger touches the square as shown in Figure E-5.10 (*a*).

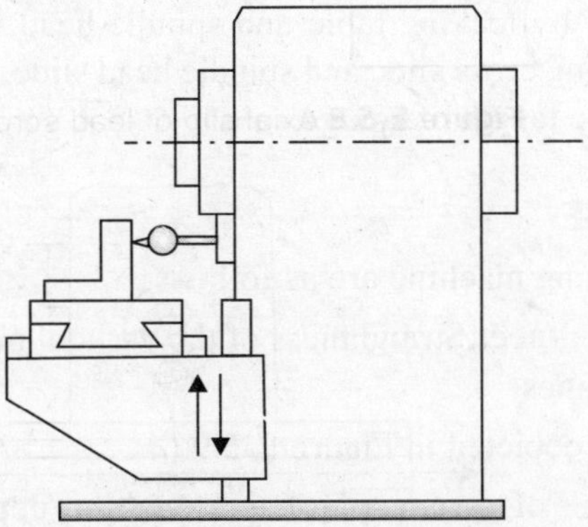

Figure E-5.10 (*a*) Straightness of the table surface to column ways for knee in the vertical plane of symmetry

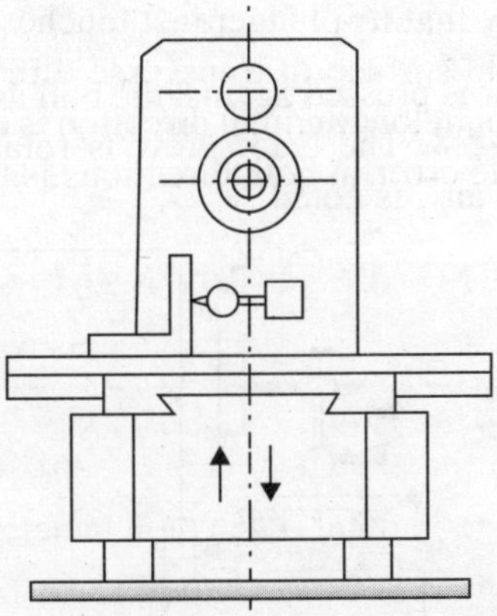

Figure E-5.10 (*b*) Straightncss of the table surface to column ways for knee in the plane perpendicular to the vertical plane

The dial indicator readings are noted at three locations *i.e.*, middle and near the extremities of the horizontal travel of the table. The difference in the readings of the dial indicator is the measure of error. The same procedure is adopted for the measurement in the perpendicular plane shown in Figure E-5.10 (*b*). The permissible error in both the planes is 0.025 mm over a length of square.

3. Flatness of the table surface: The flatness of the table surface is important requirement of the milling machine. It is checked with the help of spirit level. This test is performed bringing table and cross-slide in its central position. The table is unlocked while knee and cross-slide are locked. The spirit level is placed at several positions on the table as shown in Figure E-5.11. At all positions, the bubble of the spirit level must be in its central position. Maximum permissible error is 0.05 mm.

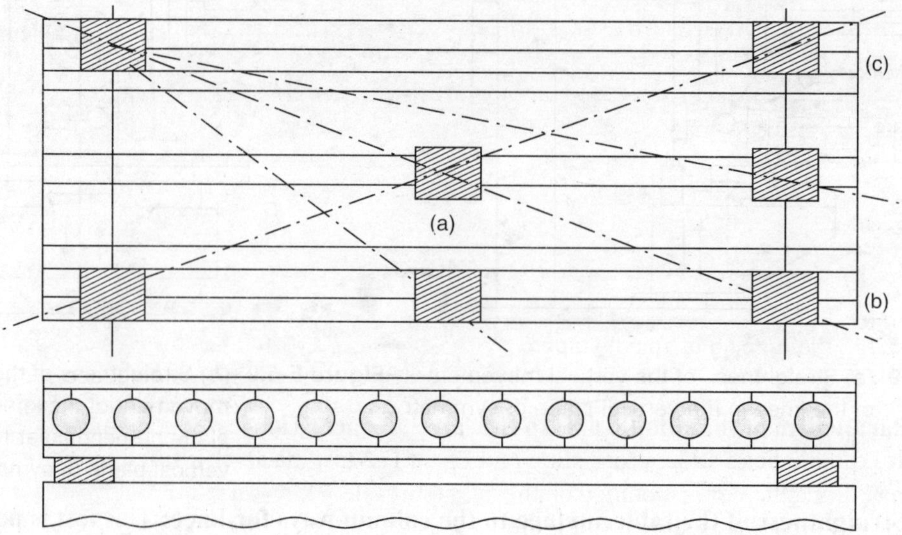

Figure E-5.11 Flatness of the table surface

4. Parallelism of the table surface to its movement: This test is performed in transverse and longitudinal directions with the help of dial indicator and straight edge. The dial indicator is mounted in such a way that its plunger just touches surface of straight edge as shown in Figure E-5.12. Parallelism of the table surface in transverse direction is checked by locking table and spindle head slide while parallelism in longitudinal direction is checked by locking cross slide and spindle head slide. Maximum permissible error in both directions is 0.02 mm over the length of square.

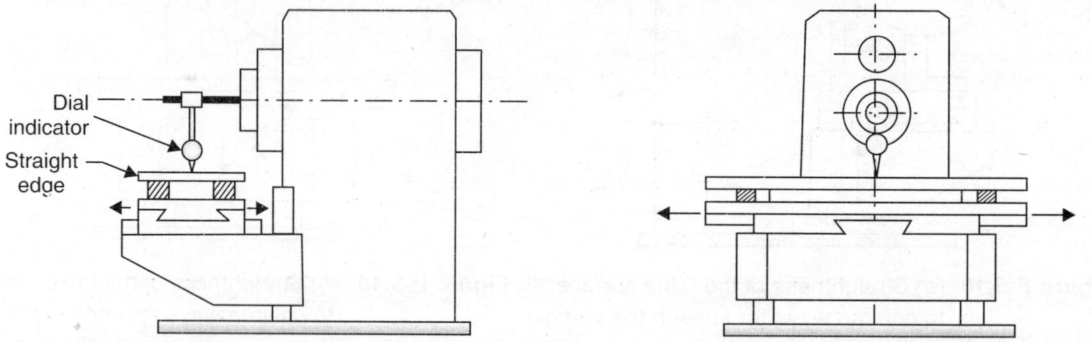

Figure E-5.12 Parallelism of the table surface to its movement in transverse and longitudinal direction

5. True running of spindle taper: This test is performed by arranging the table in its main position longitudinally and the test mandrel is fitted into the tapered hole of the spindle. The dial indicator is mounted on the table in such a way that its plunger just touches the test mandrel as shown in Figure E-5.13. The spindle is rotated slowly and readings of dial indicator are noted. The permissible error is 0.01 mm.

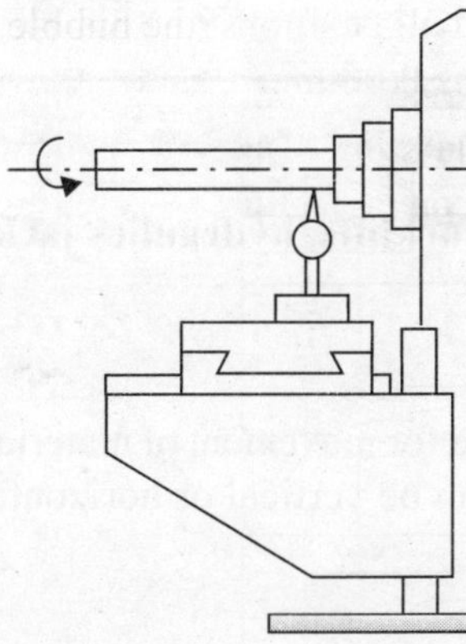

Figure E-5.13 True running of spindle taper

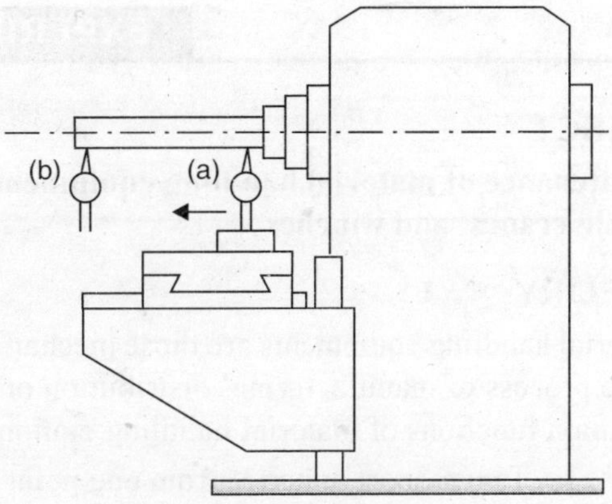

Figure E-5.14 Parallelism of the spindle axis to the table surface

6. Parallelism of the spindle axis to the table surface: This test is performed in two planes by locking table, spindle head slide, cross slide and knee. The test mandrel is fitted in the tapered hole of the spindle and dial indicator is mounted on the machine table. The plunger of dial indicator just touches the surface of test mandrel as shown in Figure E-5.14.

The readings at the maximum vertical travel of the table surface are noted at two positions. The permissible error in both planes is 0.025 mm over the length of test mandrel.

7. Squareness of the movement of the table transversely to its longitudinal movement: This test is performed by arranging the straight edge parallel to the longitudinal movement of the table. The square is then placed against the straight edge. The dial indicator is fitted as shown in Figure E-5.15. The table is locked in its central position and dial indicator readings are noted. The permissible error is 0.02 mm over the length of square.

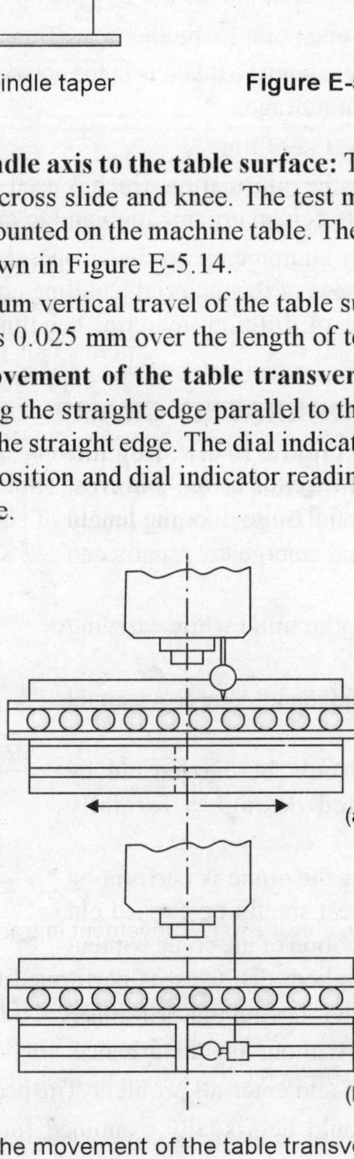

Figure E-5.15 Squareness of the movement of the table transversely to its longitudinal movement

EXPERIMENT NO. 6

OBJECT

Maintenance of material handling equipment- hand operated cranes, forklifts, hydraulics jacks, mobile cranes, and winches

THEORY

Material handling equipments are those mechanical equipments that are used for movement of materials in the process of manufacturing, distribution or disposal. The movement can be vertical or horizontal. The main functions of material handling equipments are as follows:

1. To transport material from one point to another.
2. To position material at a single point so that it is in the correct position for subsequent handling, machining, transportation or storage.
3. To store materials over a period of time.
4. To collect and communicate the information which is used to coordinate the flow of materials within organization and between an organization and to customers.

Some of the material handling equipments are hand operated crane, forklift, hydraulic jack, mobile crane and winch etc. Maintenance of the material handling equipment is very important to keep them in good condition. Maintenance of different material handling equipments is discussed in the following headings.

MAINTENANCE OF HAND OPERATED CRANE

Hand operated crane is shown in Figure E-6.1. Regular inspections should be carried out to avoid crane failures, increases safety and reduces interruptions. By performing timely maintenance, unanticipated failures and emergency repairs can be avoided.

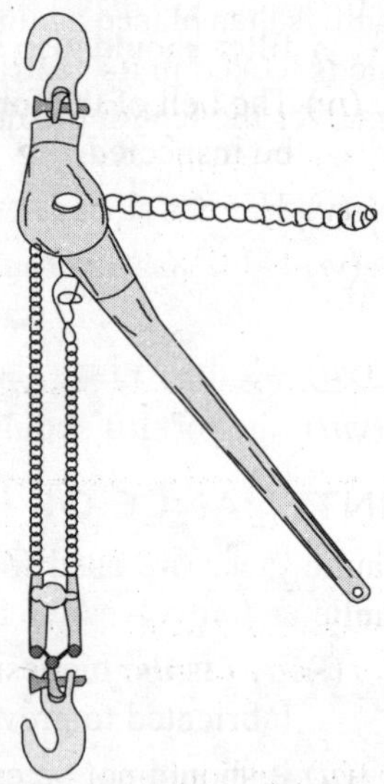

Figure E-6.1 Hand Operated Crane

Following points should be kept in mind while carrying maintenance of hand operated crane:

(*i*) Company's requirements and manufacturer's manual should be strictly followed.

(*ii*) Maintenance work should only be carried out by competent person, selected through a formally documented assessment.

(*iii*) In order to find out whether the crane is performing the intended work or not, test should be carried out to check operation of each motion of the crane without load and with rated load applied.

(*iv*) Maintenance work should be carried out at a place where it can be carried out without any disturbance.

(*v*) Use maintenance checklists and enter all problems (inspections and repairs) in the logbook.

(*vi*) The structure of crane should be visually examined for physical damage such as crack, deformation and corrosion.

(*vii*) Check for loose nuts and bolts. Check for wear of brake and clutch system.

(*viii*) Working order of all mechanical components like gear box, hydraulic system etc should be checked.

MAINTENANCE OF FORKLIFT

Forklift is shown in Figure E-6.2. It is a basically truck with additional power than a normal truck. The forklift has forks to lift and transport materials. Proper procedures should be followed while performing maintenance work on forklift.

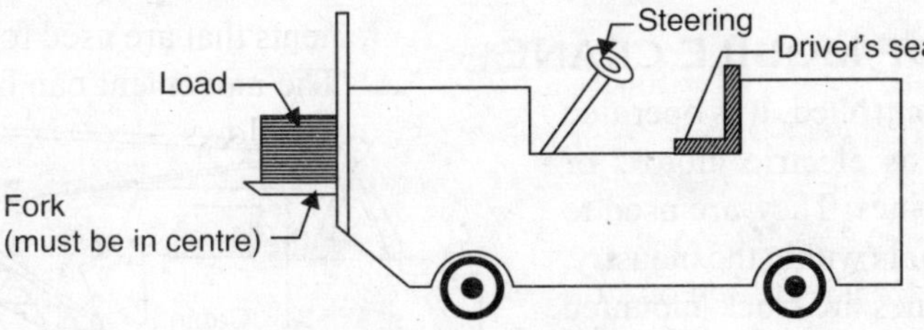

Figure E-6.2 Forklift

Maintenance of forklift involves the following steps:

(*i*) Horn, safety warning devices, cab lights, outside lights and overhead guard should be checked for proper operation.

(*ii*) Wiring and battery compartment should be checked. Battery terminal and connectors should be cleaned. Level of distilled water in the battery should be checked. Grease should be applied on the battery connectors to prevent corrosion.

(*iii*) Lubricating oil should be changed regularly as per guidelines. Oil filter, fuel filter and the air filter should also be cleaned and changed, if required.

(*iv*) The belt of the forklift engine should be checked for cracks and frays. The drive axle should be inspected.

(*v*) The clutch pedal, steering cylinder and the drag links should be checked.

(*vi*) Hoist and hydraulic lines should be inspected. The forks should be inspected for physical damage like crack, bowing and bending.

(*vii*) All the fittings should be lubricated.

(*viii*) The forklift should be wiped down and spray cleaner should be used to the forklift.

MAINTENANCE OF HYDRAULIC JACK

Hydraulic jacks are used to raise heavy items. It uses incompressible hydraulic oil. Maintenance of hydraulic jack involves the following steps:

(*i*) Dirt is the biggest cause of failure of hydraulic system. Jack should be kept clean and well lubricated to prevent foreign particle from entering the hydraulic system.

(*ii*) It should not be exposed to rain, snow, sand or dust.

(*iii*) When hydraulic jack is not in use, piston and pump rods should be kept fully retracted. Jack should be stored on its base and in area free from corrosive vapors.

(*iv*) Jack should be checked for loose hardware and leakage of hydraulic oil.

(*v*) Hydraulic jack should be filled with approved hydraulic oils only. Manufacturer's recommendation should be consulted when changing the hydraulic oil. Brake fluid should never be used. Since brake fluid contains alcohol that will damage the internal seals.

(*vi*) Jack should be checked for physical damage like cracks.

(*vii*) Hydraulic jack should be checked for excessive wear, bending or other damage. Piston rod should be checked for damage.

(*viii*) Significant damage to hydraulic jack can occur by lifting a body too heavy for its rated capacity. Exceeding the lift capacity can place significant pressure on internal seals and valves. It will not only damage jack but also result in injury to the operator. So, rated capacity of jack should be kept in mind while lifting.

MAINTENANCE OF MOBILE CRANE

Mobile crane is cable-controlled. It is operated by prime movers such as electric motors or internal combustion engines. They are used to transport different materials within the industry. Examples of mobile cranes are truck-mounted crane, rough terrain crane, all terrain crane, crawler crane and floating cranes. Figure E-6.3 shows crawler crane.

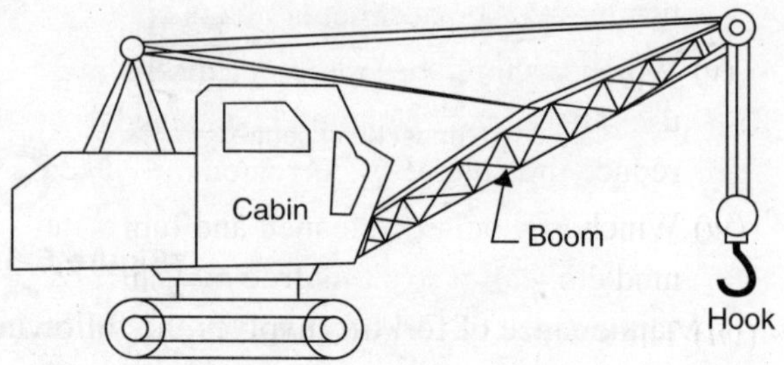

Figure E-6.3 Crawler Crane

Following points should be kept in mind while carrying maintenance of mobile crane:

(*i*) The main objective of a maintenance work is to make sure that a mobile crane is safe to be taken into, or to continue in, use. Necessary inspection is carried out as a part of the maintenance work.

(*ii*) Company's requirements and manufacturer's manual should be followed. Use maintenance checklists and enter all problems (inspections and repairs) in the logbook.

(*iii*) Check for loose nuts and bolts. Tighten them, if required.

(*iv*) Different components should be checked for wear. Also, working of all mechanical components like gear box should be checked.

(*v*) Maintenance work on mobile crane should only be carried out by competent person, selected through a formally documented assessment.

(*vi*) Mobile cranes operate in quite risky environment which includes lifting heavy loads over people. It should be taken into account by the maintenance crew when determining the scope and nature of the maintenance. Maintenance work should be carried out at a place where it can be carried out without any disturbance.

(*vii*) In order to find out whether the crane is performing the intended work or not, test should be carried out to check operation of each motion of the crane without load and with rated load applied.

(*viii*) Crane should be examined for physical damage like crack, deformation and corrosion.

MAINTENANCE OF WINCH

A winch is a mechanical device, driven by electric motor and used to pull heavy loads. Figure E-6.4 shows electric winch.

Maintenance work should be carried out by competent person as per manufacturer's guidelines. Maintenance of winch involves the following steps:

(*i*) Winches should be inspected before using. Damaged winches and inoperable winches should be discarded immediately to avoid chances of accident.

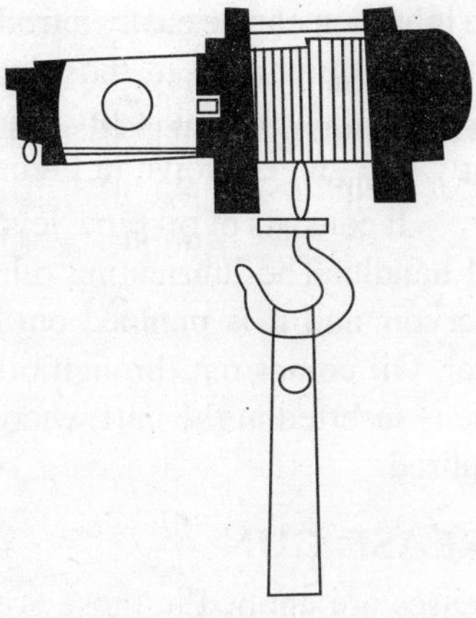

Figure E-6.4 Electric Winch

(ii) Cable shouid be inspected for frays, cuts and abrasions. Inspection should be carried out before and after each use. If the cable is beyond repair, it should be immediately replaced by a new cable.

(iii) Cable should be washed after riding through mud or saltwater.

(iv) The working of mechanical component should be inspected. If the motor is running but drum is not turning, this means that clutch is not working properly.

(v) Winches should be stored in a dry location when not in use. Dirt in winch should be cleaned before storage to reduce the chance of corrosion.

(vi) Winches should be cleaned and lubricated. Dirt, dust, mud etc. can restrict its free movement.

(vii) The solenoid or remote control switch should be checked. If the motor is running in only one direction, the solenoid is defective or the remote control switch is inoperative.

(viii) Winches should never be overloaded. As overloading the cable can lead to failure.

EXPERIMENT NO. 7

OBJECT

Use of lubrication equipments like oil gun, grease gun

INTRODUCTION

The science which deals with reduction of friction by application of lubricant between the rubbing surfaces of bodies having relative motion is known as lubrication. Lubrication is essential for correct operation of mechanical systems like pistons, pumps, cams, bearings, turbines, cutting tools etc, as without lubrication the pressure between the surfaces in close proximity generates enough heat which rapidly causes surface damage, in a coarsened condition which may cause seizure literally by welding the surfaces together. Lubrication performs the following major functions:

(i) Minimization of the frictional power loss.

(ii) Minimization of the wear of the rubbing parts.

(iii) Minimization of rusting and corrosion.

(iv) It works as coolant.

(v) It also keeps machine parts clean and free from dirt.

Commonly used lubrication equipments in workshop and garage are:

(i) Oil gun

(ii) Grease gun

OIL GUN

In workshops, lubrication of various machine parts is carried by means of lubrication equipments like oil gun, grease gun etc. These equipments can be used for both external as well as internal lubrication.

The lubricant can be easily introduced by means of these equipments to those interior parts of machines, where the operator cannot reach easily. Oil gun is shown in Figure E-7.1.

It consists of oil pipe, lever, oil reservoir and handle. The lubricating oil is filled in oil reservoir and it is pumped out by the help of lever. Oil comes out through oil pipe. This oil pipe is inserted at the part where lubrication is required.

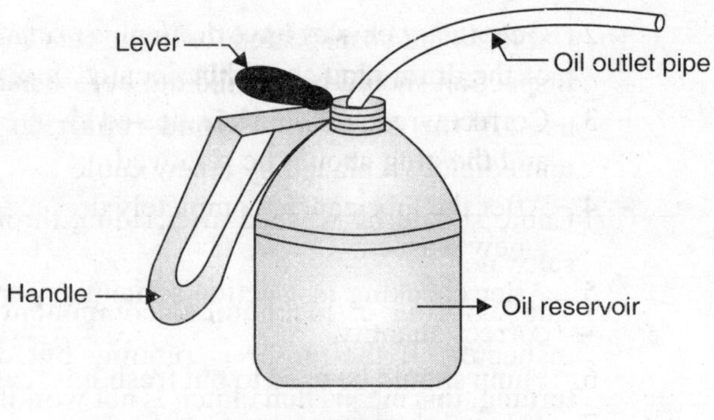

Figure E-7.1 Oil Gun

GREASE GUN

Greases are applied in those areas where a continuous supply of oil cannot be maintained, such as in open bearings or gears. Grease is the most commonly used lubricant in roller bearings and low velocity applications. Grease gun is shown in Figure E-7.2.

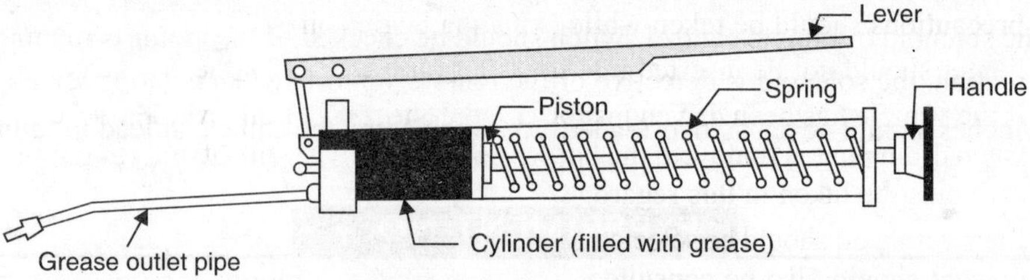

Figure E-7.2 Grease Gun

It consists of cylinder, piston, spring, lever etc. The grease is filled in the cylinder and piston reciprocates inside the cylinder. Piston develops low pressure which forces the grease out of the gun. Manual grease guns sometimes have disadvantages of over and under lubrication. Technician should know the amount of the grease coming out per stroke of the grease gun. The different types of grease guns available usually are lever type, pistol-grip type, hand grip type and air-powered etc.

EXPERIMENT NO. 8

OBJECT

Removing old lubricant, cleaning and replenishing machine with fresh lubricant

INTRODUCTION

Proper lubrication with timely replacement plays a significant role in maintaining the accuracy of the machine and enhances life of the machine. The process of lubrication, cleaning and replenishing machine with fresh lubricant should be carried out by trained person. The machine instruction's manual should be consulted for lubrication instructions. These manual have not only details of lubrication schedule but also specification of lubricant to be used.

PROCEDURE OF CHANGING ENGINE OIL

The general instructions of changing engine oil to be adopted are as follows:

1. The vehicle should be raised and it should be ensured that the vehicle is in level and stable. Leakage should be checked visually.

2. Generally vehicles have drain plugs fitted at the place which is easily accessible. The position of the drain plug should be located and guard covers, if any, should be removed.
3. Correct wrench should be selected to loosen the plug. A large diameter pan should be positioned and the plug should be removed.
4. After the lubricant is completely drained out, drain plug should be cleaned and put back with a new washer.
5. After checking instruction's manual, correct specification lubricating oil should be used in correct quantity.
6. Pump should be used to put fresh lubricating oil into the filler hole or dip-stick tube opening.
7. After each pump, a pause of few seconds should be given so that the lubricating oil flows to the bottom.
8. When the lubricating oil has reached to the bottom of the filler plug hole, or to the top of the mark on the dip-stick, pumping should be stopped.

PRECAUTIONS

Following precautions should be taken while changing engine oil:

(*i*) Finger should not be inserted into the filler plug hole when the engine is running.

(*ii*) If the vehicle has been driven just before the checking of oil level, the oil could be hot. Thus, when drain plug is removed, the hot lubricating oil may run out and result in burning. Hence, care must be taken in this regard.

(*iii*) Lubricating oil should be selected as per recommendations of the manufacturer. The instruction manual should also be consulted. Selection of wrong lubricating oil can result in machine failure.

(*iv*) A correct oil level is important. If the level of oil is too high, the oil may churn and aerate, causing bubbles that can trap heat and allow the oil to run hotter. This may cause early failure of bearings and gears. On the other hand, if the level of oil is too low, air may enter the system and heat generation can result in the burning of components.

EXPERIMENT NO. 9

OBJECT

Case study of computerized maintenance schedule

COMPUTERIZATION OF MAINTENANCE

In today's techno-savvy world, quick and economical communication, job preparation, information collection and maintenance are accomplished far more efficiently by means of computer. Computerization of maintenance is required for maintaining database of information about different maintenance operations of an organization. This information can be used by maintenance workers to carry out their duties in a more effective way. It is used to store records such as information related to machines that require maintenance and storerooms containing the required tools/ spare parts. Computerization of maintenance also helps in planning suitable maintenance program, tracking repair work, taking quick decisions, conducting cost analysis and verification of regulatory compliance. Such information can also be used when dealing with third parties, like in a liability case where the information can serve as a evidence that proper maintenance had been carried out.

COMPUTERIZED MAINTENANCE MANAGEMENT SYSTEM (CMMS)

The term used for computerized maintenance system is Computerized Maintenance Management Systems (CMMS). However, Computerized Maintenance Management Information Systems (CMMIS) is more preferred since today's systems by design and by use are not used to manage maintenance but rather to inform about maintenance. CMMIS allows easier access to large amounts of information reducing time and cost.

CMMIS is used to perform preventive maintenance, asset management, work order and inventory control. It can also produce status reports and documents giving details of maintenance work. CMMIS take care of scheduling jobs, assigning duties, materials management, spare parts management and tracking information related to the cause of the problem. It also maintains data about machines like specifications, warranty and service contracts. It maintains different important documents required for the processing of safety requirements. Thus, perform various functions and offers several advantages

The above mentioned functions are formed by softwares available in market. These softwares can be either web-based or LAN Based. With the help of these softwares, maintenance work can be performed in effective and efficient manner. CMMS/ CMMIS software must have various characteristics such as it must be a sound, comprehensive, on-line and user-friendly. The CMMIS software offers the following benefits:

1. Improves speed and accuracy of the process
2. Improves efficiency of maintenance resources and hence, lowers unit cost
3. Helps in proper utilization of resources
4. Provides better performance and product quality
5. Improves service to the customers
6. Improves asset reliability, capacity assurance and equipment uptime

CASE STUDY REPORT

Students should be asked to conduct case study and prepare repart on Computerized Maintenance schedule of any industries.

EXPERIMENT NO. 10

OBJECT

Reconditioning of machine parts

MAINTENANCE OF MACHINE PARTS

Maintenance of various machine parts like belt drive, chain drive, gear drive, shaft coupling and threaded fasteners is below.

MAINTENANCE OF BELT DRIVE

Belt is one of the most common and effective device used for transmission of motion and power from one shaft to the other by means of a thin inextensible band (belt) running over pulleys. Figure E-10.1 shows belt drive.

A well-designed and maintained belt drive is capable of operating for many years when used under normal working conditions. The belt drives, which operate at high speeds, heavy loads, frequent stop/start conditions and at high working temperature require special attention. Some of the points related to maintenance of belt drive are as follows:

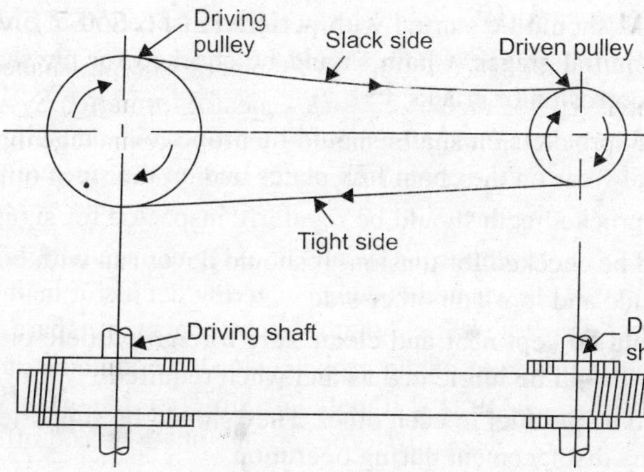

Figure E-10.1 Belt Drive

1. Maintenance work should be started with periodic inspection. The alignment and tension of belt should be carefully checked.
2. Belt drive should be inspected for wear or rubbing against drive components.
3. Belt should also be checked for damage. It should be free from oil, grease, debris and dust. Any accumulation on belts can damage it.
4. Drive components such as bearings, shafts, motor also requires frequent inspection.
5. The operating speed of belt drive should also be checked regularly. The working temperature also affects the life of belt. Higher working temperature reduces life of the belt.
6. Belt drive should be inspected for abnormal sound and mechanical vibrations. A properly maintained belt drive will operate smoothly and quietly without any vibrations.
7. Belt drive guard should not be loose or damaged.
8. The working environment in and around the belt drive should be safe. The working area should be free from debris and obstructions. Floors should be neat and clean.

MAINTENANCE OF CHAIN DRIVE

Chain drive is shown in Figure E-10.2. It is used for transmission of power from shaft to the other by means of chain and sprockets. It is used in cycles, motorcycles and many machines.

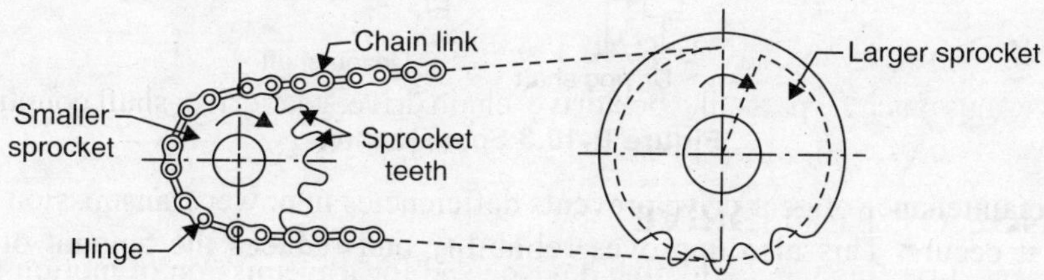

Figure E-10.2 Chain Drive

Like other mechanical devices, chain drive also requires maintenance in order to obtain satisfactory working. The following instructions should be followed regarding maintenance of chain drive:

1. Maintenance work should be started with periodic inspection in order to identify potential problems during initial stages. Chain should be checked for physical damage like broken part, loose pins, corrosion or cracks.
2. The alignment of sprockets on shafts should be proper. Any misalignment of sprockets can result in abnormal wear on the chain link plates and on the sides of the sprocket teeth.
3. The side of the sprocket teeth should be regularly inspected for signs of wear.
4. The chain should be checked for tension. It should never run with both sides tight. It should be tight on one side and loose on other side.
5. Chain drive should be kept neat and clean. Any foreign particle on the chain can result in abrasive wear. It should be lubricated as and when required.
6. The shafts should be parallel to each other. They should be supported by strong bearings in order to avoid any displacement during operation.
7. The sprockets should be installed on the shaft according to the manufacturer's guideline.
8. Chain should be inspected for flexibility and elongation.

MAINTENANCE OF GEAR DRIVE

Gears are basically toothed wheels, which are used to transmit power and motion from one shaft to another by means of successive engagement of teeth. They can be used to drive shafts that are parallel, intersecting or neither parallel non-intersecting. Preference to gear drive is given over belt drives and chain drives, where moderate or large amount of power has to be transmitted at a constant velocity ratio. Different types of gears available are: spur gear, helical gear, bevel gear and worm gear. Spur gearing is shown in Figure E-10.3.

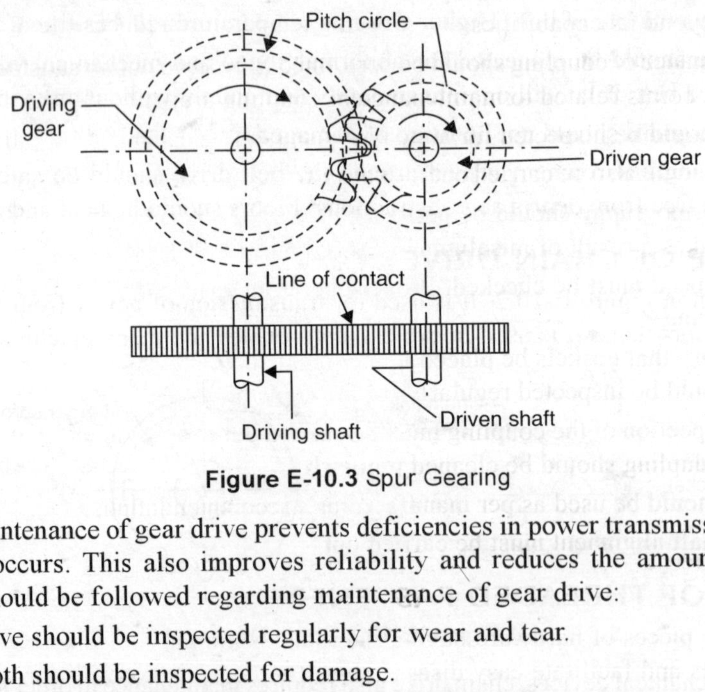

Figure E-10.3 Spur Gearing

Regular maintenance of gear drive prevents deficiencies in power transmission and prevents the damage before it occurs. This also improves reliability and reduces the amount of downtime. The following points should be followed regarding maintenance of gear drive:

1. Gear drive should be inspected regularly for wear and tear.
2. Gear tooth should be inspected for damage.
3. Gear should be checked for physical damage like crack or corrosion.
4. Gear must be inspected for surface damage like pitting. Gear must be grinded and polished properly in order to avoid pitting, which usually occurs in initial stages of operation.

5. They should be properly and regularly lubricated with proper gear oil. This will protect them from corrosion.
6. The lubricant should be used as per manufacturer's recommendations.
7. The gear should be regularly cleaned using various washing liquids. Regular cleaning of the gear removes contaminants and maintains their performance. Cleaning also enhances life of the gear.
8. Gear drive should be checked for abnormal sound and excessive mechanical vibrations.

MAINTENANCE OF COUPLING

Couplings are used to connect two shafts together at their ends for the purpose of transmitting power from one shaft to another. It also provides for shaft misalignment and introduces mechanical flexibility. The various couplings are:

(*i*) Rigid Coupling
- (*a*) Sleeve Coupling
- (*b*) Flange Coupling

(*ii*) Flexible Coupling
- (*a*) Oldham Coupling
- (*b*) Universal Coupling
- (*c*) Elastomeric Coupling

Figure E-10.4 shows flange coupling. The various reasons for coupling failure are improper installation of coupling, poor coupling selection, operating beyond its capabilities.

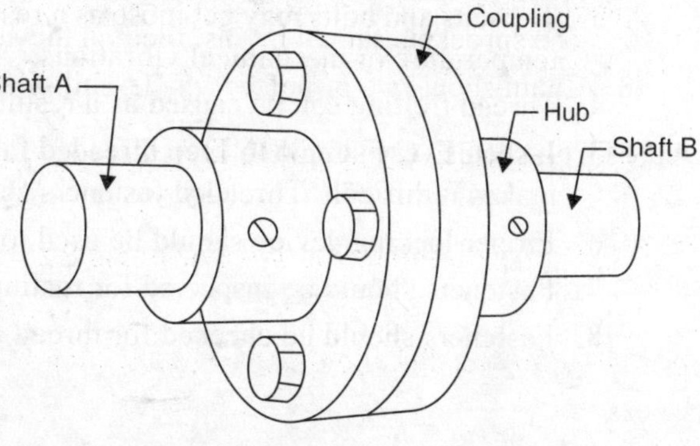

Figure E-10.4 Flange Coupling

Regular maintenance of coupling should be performed in order to obtain satisfactory performance. Some of the important points related to maintenance of coupling are:

1. Coupling should be inspected for abnormal sound and mechanical vibrations.
2. Inspection should also be carried out to check excessive heat generation and lubricant leakage.
3. Infrared thermography should be performed to measure the heat of the coupling. Heat may be generated as a result of misalignment, axial movement or torsion.
4. Operating speed must be checked. Proper operating speed should be maintained for longer life of coupling.
5. It is necessary that gaskets be placed in the female coupling so tight connection is achieved. Gaskets should be inspected regularly for the signs of wear or cuts.
6. Visually inspection of the coupling must be carried out in order to check wear, corrosion and cracking. Coupling should be cleaned regularly.
7. Lubricant should be used as per manufacturer's recommendations.
8. Precision shaft alignment must be carried out.

MAINTENANCE OF THREADED FASTENERS

Threaded fasteners are pieces of hardware having internal or external threads. They are employed to assemble different parts and facilitate easy disassembly. The common types of threaded fasteners are screw, nut, stud and bolt. An external thread is a thread on the outside of a shaft. The threads on a bolt are external threads. An internal thread is a thread on the inside of a hole. The threads in a nut are internal threads. Nut and bolt are shown in Figure E-10.5.

The following points should be remembered regarding maintenance of threaded fasteners:

1. There are three modes of threaded fastener failure. Failures through the shank or threaded section of the fastener, external thread damage and internal thread damage. Fasteners should be properly installed as improper installation can damage threads. Use of incorrect fastener size can also damage thread.
2. Threaded fastener should be checked for wear as the wear of fastener material can take place as a result of repeated use.

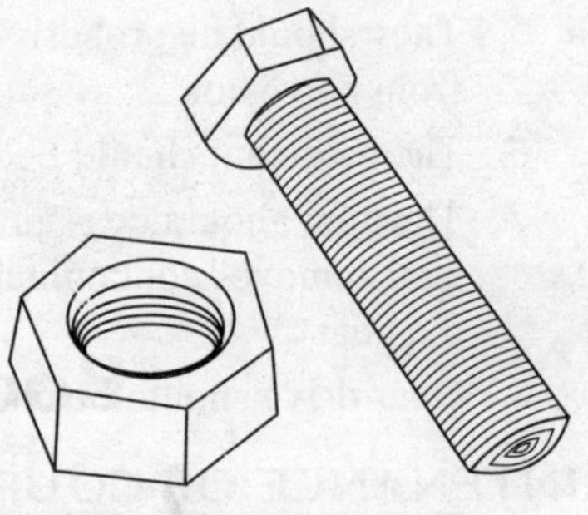

Figure E-10.5 Nut and bolt

3. The nuts and bolts may get loose as a result of operating in conditions such as varying stress, temperature or mechanical vibrations.
4. Thread fretting can be caused as a result of excessive vibrations.
5. It seems very simple to keep threaded fasteners tight, but the moving nature of the machines makes it difficult. Threaded fasteners should be checked whether they are tight or not.
6. Proper locking device should be used to prevent the loosening of fasteners.
7. Fasteners should be inspected for rusting.
8. Fasteners should be checked for thread galling.

Books Referred

1. T. R. Banga, N.K. Agarwal and S.C. Sharma, *Industrial Engineering and Management Science*, Khanna Publishers, Delhi, 2004.
2. A.P. Verma, *Industrial Engineering and Management*, S.K. Kataria and Sons, Delhi, 2009.
3. J.S. Narang, S.K. Walia and A. Gupta, *Installation, Testing and Maintenance*, Dhanpat Rai and Co. (P) Ltd., Delhi, 2006.
4. H.P. Garg and B.P. Gupta, *Industrial Maintenance,* S. Chand & Sons Co. Ltd., New Delhi, 2009.
5. M. Mahajan, *A Text-Book of Metrology*, Dhanpat Rai and Co. (P) Ltd., Delhi, 2006.
6. R.K. Jain, *Engineering Metrology*, Khanna Publishers, Delhi, 2007.
7. B.S. Raghuwanshi, *A Course in Workshop Technology Vol. II Machine Tools*, Dhanpat Rai and Co. (P) Ltd., Delhi, 2008.
8. K.C. Pal, *Resource Materials on Installation and Maintenance of Equipment*, Technical Teacher's Training Institute, Calcutta, 1984.
9. T.G. Hicks, *Pump Operation and Maintenance*, Tata McGraw Hill, New Delhi, 2008.
10. S.S. Dara, *A Text Book in Engineering Chemistry*, S. Chand & Company Ltd., New Delhi, 1992.
11. S. Trymbaka Murthy, *A Text Book of Elements of Mechanical Engineering*, I.K. International Publishing House Pvt. Ltd., New Delhi, 2006.
12. S.K. Srivastava, *Maintenance Engineering and Management*, S. Chand & Company Ltd., New Delhi, 2009.
13. Proceedings of the International Conference on Recent Trends in Engineering, Technology and Management (Editors Dr. Abhai Kumar Verma and Ashish Gupta), Excel India Publishers, New Delhi. 2011.
14. S.N. Bhattacharya, *Installation, Servicing and Maintenance*, S. Chand and Company Ltd., New Delhi, 2008.
15. I.P. Singh, *Material Science,* Jain Brothers, New Delhi, 2000.

Index

A

B

C

D

E

Win Prizes

ATTENTION READERS

Win Prizes

We request for your frank assessment regarding various aspects of the book as given below:

Installation, Testing and Maintenance

BY : Mohammad Yunus Khan

Please Fill up the Sheet in Neat and Clean Writing

(i) What topics of your syllabus, important for your examination point of view, are not covered in this book ?

....

....

....

....

(ii) What are the chapters wherein the treatment of the subject matter is not systematic or organised or updated ?

....

....

....

....

(iii) Have you come across misprints/mistakes/factual inaccuracies in the book ? Please specify the chapters and the page numbers.

....

....

....

....

(iv) What other books on the same subject have you found/heard better than the present book ? Please specify in terms of price and quality.

....

....

....

....

(v) Further suggestions and comments for the improvement of this book.

....

....

....

....

(PS : If need be please attach a separate sheet to write your views)

OTHER DETAILS

(i) Who recommended you this book ? Please tick (✓) in the box near the option relevant to you.

☐ Teacher ☐ Friends ☐ Bookseller

....

....

....

....

(ii) Name of the recommending teacher, his designation and address:

....

....

....

....

(iii) Name and address of the bookseller you purchased the book from :

....

....

....

....

(iv) Name and address of your institution (Please mention the University or Board as the case may be)

....

....

....

....

(v) Your name and complete address :

....

....

....

....

(vi) Course you are enrolled in :

....

....

....

....

The best assessment will be awarded each month : The award will be in the form of our publications as decided by the Editorial Board.

Please mail the filled up coupon at your earliest to :

S.K. KATARIA & SONS®
109, Prakash Mahal, Dr. Subhash Bhargav Lane,
Opposite Delhi Medical Association, Daryaganj, New Delhi-110002
☎ 011-23243489, 23269324; Telefax: 011-23243489
e-mail: katariabooks@yahoo.com; katariabook@gmail.com
Website : www.skkatariaandsons.com